Trees: Their Natural History
Peter Thomas

Trees are familiar, and often beautiful, components of many landscapes, vital to the healthy functioning of the global ecosystem and unparalleled in the range of materials that they provide for human use. Yet how much do we really understand about how they work? This book provides a comprehensive introduction to the natural history of trees, presenting information on all aspects of tree biology and ecology in an easy-to-read and concise text. Fascinating insights into the workings of these everyday plants are uncovered throughout the book, with questions such as how trees are designed, how they grow and reproduce, and why they eventually die, tackled in an illuminating way. Written for a non-technical audience, the book is none the less rigorous in its treatment and will therefore provide a valuable source of reference for beginning students as well as those with a less formal interest in this fascinating group of plants.

Peter Thomas is a lecturer in environmental science at Keele University, UK, where his teaching encompasses a wide range of tree-related topics including wood structure and identification, tree design and biomechanics, tree ecology and identification, and woodland management. His research interests focus on the reconstruction of past environments from tree rings, the role of trees in nature conservation and the interaction of fire with trees.

Trees: Their Natural History

Peter Thomas

PUBLISHED BY THE PRESS SYNDICATE OF THE UNIVERSITY OF CAMBRIDGE
The Pitt Building, Trumpington Street, Cambridge, United Kingdom

CAMBRIDGE UNIVERSITY PRESS
The Edinburgh Building, Cambridge CB2 2RU, UK
40 West 20th Street, New York, NY 10011–4211, USA
10 Stamford Road, Oakleigh, VIC 3166, Australia
Ruiz de Alarćon 13, 28014 Madrid, Spain
Dock House, The Waterfront, Cape Town 8001, South Africa

http://www.cambridge.org

First published 2000
Reprinted 2000 (twice)

Printed in the United Kingdom at the University Press, Cambridge

Typeface Minion 10/12½pt. System QuarkXPress™ [wv]

A catalogue record for this book is available from the British Library

Library of Congress Cataloguing in Publication data

Thomas, Peter, 1957–
Trees: their natural history/Peter Thomas
p. cm.
Includes bibliographical references (p.).
ISBN 0 521 45351 8. – ISBN 0 521 45963 X (pbk.)
1. Trees. I. Title.
QK475.T58 2000
582.16–DC21 99-15473 CIP

ISBN 0 521 45351 8 hardback
ISBN 0 521 45963 X paperback

But blessed is the man who trusts in the Lord, whose confidence is in him.
He will be like a tree planted by the water that sends out its roots by the stream.
It does not fear when heat comes; its leaves are always green.
It has no worries in a year of drought and never fails to bear fruit.

Jeremiah 17:7–8

Contents

Preface

Why write a book on trees? The motivation for me came from the frustration of trying to teach a subject where much is known but is scattered over a huge range of journals and books from many countries. There are so many fascinating stories to be told about the ways in which trees cope with the world and the problems of being large and long-lived: they are extremely well designed. *Someone* had to write this book!

My goal was to draw together strands of information to create a readable book that would answer common questions about trees, set right a number of myths and open up the remarkable world of how trees work, grow, reproduce and die. It is for you, the reader, to judge whether I have been successful. Please let me know where you find errors or would wish to argue with the logic.

I am indebted to all those who helped with this book, especially Roger Davidson and Bill Williams who read and commented on the whole manuscript and numerous colleagues who commented on parts. Val Brown, P.B. Tomlinson, Colin Black and K.J. Niklas kindly provided detailed information. Maria Murphy and Lynn Davy at Cambridge University Press are to be congratulated for their extreme patience with a faltering author. And, of course, my wife and sons are thanked for putting up with my pet project.

The following are gratefully acknowledged for their help with the diagrams: Ian Wright, Lee Manby (who drew Figures 4.2 and 7.11–7.14) and John Stanley (for drawing Figure 3.7). Crown copyright material is reproduced with the permission of the Controller of Her Majesty's Stationery Office.

Peter Thomas
Keele
May 1999

Chapter 1: An overview

What is a tree?

Everyone knows what a tree is: a large woody thing that provides shade. Oaks, pines and similarly large majestic trees probably come immediately to mind. A stricter, botanical definition is that a tree is any plant with a self-supporting, perennial woody stem (i.e. living for more than one year). The first question that normally comes back at this point is to ask what then is a shrub? To horticulturalists, a 'tree' is defined as having a single stem more than 6 m (20 ft) tall, which branches at some distance above ground, whereas a shrub has multiple stems from the ground and is less than 6 m tall. This is a convenient definition for those writing tree identification books who wish to limit the number of species they must include. In this book, however, shrubs are thought of as being just small trees since they work in exactly the same way as their bigger neighbours. Thus, 'trees' cover the towering giants over 100 m through to the little sprawling alpine willows no more than a few centimetres tall.

Some plants can be clearly excluded from the tree definition. Lianas and other climbers are not self-supporting (although some examples are included in this book), and those plants with woody stems that die down to the ground each year, such as asparagus, do not have a perennial woody stem. Bananas are not trees because they have no wood (the trunk is made from leaf stalks squeezed together). Nor are bamboos trees since they are just hardened grasses even though they can be up to 25 m tall and 25 cm thick (see Box 1.1).

An interesting feature of trees is how unrelated they are. It is generally easy to say whether a plant is an orchid or not because all orchids belong to the same family and share a similarity in structure (especially the flowers). This is true of most plant groups such as grasses and cacti (in their own families) and chrysanthemums (all in the same genus). But the tree habit has evolved in a wide range of plants so there is no other similarity other than being tall with a perennial skeleton. Box 1.1 illustrates how many groups have evolved the tree habit. This is a superb example of 'convergent evolution' where a number of unrelated types of plant have evolved the same answer—height—to the same problem: how to get a good supply of light.

On the whole, this book is concerned with the two biggest groups of trees. These are the **conifers** and their allies, and the **hardwoods** like oak, birch and so on. (As you can see from Box 1.2 the terminology can be confusing, so throughout

Box 1.1 The range of trees found in different plant groups

Ferns (Pteridophytes)	Tree ferns: all in the family Cyatheaceae; rarely branched, no true bark and with a trunk containing woody strands; need frost-free shaded habitats
Conifers and their allies: Gymnosperms. This term means 'naked seeds' (as in gymnasium, where the Greeks exercised naked); the seeds are exposed to the air and can be seen in the cone or fruit without having to cut anything open	Conifers: 600+ species in three families Cupressaceae: cypress, junipers Taxodiaceae: redwoods Pinaceae: pines, spruces, larches, hemlocks, firs, cedars Taxads: *c.* 20 species Taxaceae: yews Ginkgo: 1 species Ginkgoaceae: the ginkgo or maidenhair tree (*Ginkgo biloba*) Cycads: palm-like with stiff leathery leaves Gnetales: a strange group with a few interesting woody plants *Welwitschia mirabilis*: single species in SW Africa *Gnetum*: mostly tropical climbers *Ephedra*: 30+ low shrubs of dry deserts
Flowering plants: Angiosperms. This means hidden seeds: contained inside a fruit	**Dicotyledons** The main group of trees such as oaks, birches, etc. Around 75 of the world's 180 families contain trees **Monocotyledons** A wide-ranging set of trees concentrated in a few families Palmaceae: palms; mostly tropical, a few temperate; nearly 3000 species Agavaceae: dragon trees (*Dracaena* spp.); mostly N. African Pandanaceae: screw pines (*Pandanus* spp.); Old World Tropics. Stilt roots supporting a stout forked trunk

Box 1.1 *(cont.)*

	Liliaceae: aloes (*Aloe* spp.); Southern Africa yuccas (including the Joshua tree, *Yucca brevifolia*) cordyline palms (*Cordyline* spp.); Australia and New Zealand European butcher's brooms (*Ruscus* spp.) Xanthorrhoeaceae: grass trees (*Xanthorrhoea* spp.); Australia. Short trunk with forked branches and long narrow leaves Strelitziaceae: traveller's palm (*Ravenala madagascariensis*) **Monocotyledons that are not trees** Musaceae: bananas (*Musa* spp.). The trunk is made from leaf stalks squeezed together Poaceae: bamboos (e.g. *Dendrocalamus* spp.) are just hardened grasses with no wood

this book we will stick to conifers and hardwoods as shorthand for gymnosperms and dicotyledon angiosperms.) The monocotyledon trees such as palms and dragon trees are mentioned in passing but on the whole they grow in a different way from conifers and hardwoods and the book can only be so long. Purists might indeed argue that since the trunks of these trees contain no real 'wood' (Chapter 3) they are not trees anyway. Tree ferns (Box 1.1) come into the same category.

A short history of trees

Back in the Silurian, over 400 million years ago, the first vascular plants (those with internal plumbing) appeared on the Earth. From these the first trees evolved in the early Devonian around 390 million years ago. Within 100 million years, the coal-producing swamps of the Carboniferous (360–290 million years ago) were dominated by lush forests. We would have recognised the tree ferns from today's forests but the others—giant horsetails and clubmosses—have long since disappeared leaving us just a few small relatives. The horsetails such as *Calamites* were up to 9 m tall and 30 cm in diameter but the clubmosses (notably *Lepidodendron*) must have been magnificent at up to 40 m high and a metre in diameter. In these forests the first primitive conifers appeared and by around 250 million years ago (the late Permian) trees such as cycads, ginkgos and monkey

Box 1.2 Definitions that go with the two main groups of trees

Throughout this book the terms **Conifers** and **Hardwoods** will be used as shorthand for Gymnosperms and Angiosperm trees.

Gymnosperms	**Angiosperms**

As explained in Box 1.1, these are the proper botanical terms but a little hard to digest.

Conifers and their allies	The rest

As you can see from Box 1.1 the gymnosperms include more than just the conifers but they are the major component.

Softwoods	**Hardwoods**

The problem with these descriptive terms (which stem from the timber industry) is that although most gymnosperms *do* produce softer wood, there are many exceptions, and many hardwoods can be physically soft. Yew (*Taxus baccata*, a softwood) produces very dense and hard wood, whereas some Hardwoods, like balsa (*Ochroma pyramidale*), are very soft and easily broken or indented with a fingernail.

Evergreens	**Deciduous trees**

Exceptions can be found here as well. The dawn redwood (*Metasequoia glyptostroboides*), the swamp cypress (*Taxodium distichum*) and larches (*Larix* species), for example, are deciduous gymnosperms. In contrast, European holly (*Ilex aquifoilium*), rhododendrons and many tropical angiosperms are evergreen.

Needle trees	**Broadleaved trees**

Most conifers indeed have needle-shaped leaves but again there are exceptions. The ginkgo tree (*Ginkgo biloba*) and monkey puzzle (*Araucaria araucana*) have definite broad flat leaves (admittedly these trees are easily identified oddities). Cycads, which are primitive gymnosperms, have long divided leaves that resemble those of palms. Some angiosperms have reverted to needle leaves or have largely lost their leaves and use their needle-like branches as leaves, e.g. gorses (*Ulex* spp.) and brooms (*Cytisus* spp.).

puzzles were recognisable: the sort of trees found fossilised in the petrified forest of Arizona. The pines were not far behind, probably evolving around 180–135 million years ago (Jurassic) to share the earth with the dinosaurs.

Conifer domination was long and illustrious but the early hardwoods were diversifying during the early Cretaceous, around 120 million years ago. The hardwoods probably evolved from a now extinct conifer group that had insect-pollinated cones. The magnolias are some of the earliest types of hardwood that we still have around. During the Cretaceous period and into the early Tertiary (65–25 million years ago) the hardwoods underwent a massive expansion, displacing the conifers, undoubtedly helped by the warm humid global climate of the early Tertiary.

At the end of the Permian period, around 250 million years ago, most of the Earth's land masses were squashed together into the old super-continent of Pangaea. By the time the hardwoods had evolved, Pangaea had broken into Laurasia (which gave rise to the northern hemisphere continents) and Gondwanaland (containing what is now Australia, Africa, South America, India and Antarctica), trapping the pines primarily in the northern hemisphere. Laurasia and Gondwanaland themselves broke apart later, which goes some way to explaining why the hardwoods of the northern and southern hemisphere are so different from each other and yet remarkably similar around the globe within a hemisphere.

By 95 million years ago (midway through the Cretaceous period) a number of trees we would recognise today were around: laurels, magnolias, planes, maples, oaks, willows and, within another 20 million years, the palms. When the dinosaurs were disappearing (by 65 million years ago) the hardwoods were dominating the world with the conifers exiled mostly into the high latitudes.

Living fossils

Most of the types of tree we see every day have been around for a long time. Perhaps the most incredible are the growing number of rediscovered 'living fossils': trees known from the fossil record and which were thought to have become extinct and yet have been found hanging on in remote parts of the world. The most famous is the ginkgo (Japanese for 'silver apricot', named after the fruit) or maidenhair tree (*Ginkgo biloba*), a Chinese tree known from the fossil record back about 180 million years—the Jurassic era—and 'rediscovered' in Japan by Europeans in 1690. The dawn redwood (*Metasequoia glyptostroboides*) was similarly refound in 1945 in China, and more recently in 1994 the Wollemi pine (*Wollemia nobilis*, a member of the monkey puzzle family, Araucariaceae) was found growing in Wollemi National Park near Sydney, Australia. In these plants you can see real history (or prehistory) and touch plants that would have been familiar to the dinosaurs!

The value of trees

Over their long history, trees have played an important part in our lives. Trees have been (and still are) sacred to many peoples; oaks were sacred to the European Druids, baobabs (*Adansonia digitata*) to African tribes, the ginkgo (*Ginkgo biloba*) to the Chinese and Japanese, sequoias to North American Indians, and monkey puzzles to the Pehuenche people of Chile. Indeed, many of our words and expressions are derived from a close association with trees. Writing tablets were once made from slivers of beech wood (*Fagus sylvatica*), hence 'beech' is the Anglo-Saxon word for book. Beech is still called 'bok' in Swedish and 'beuk' in Danish. Romans crowned athletes with wreaths of the bay laurel (*Laurus nobilis*); this was extended to poets and scholars in Middle Ages, hence Poet Laureate. Similarly, Roman students were called bachelors from the laurel berry (baccalaureus) leaving us with bachelor degrees (baccalaureate) and, since Roman students were forbidden to marry, unmarried bachelor males.

Despite modern technology, we are still very reliant on wood as a raw material. The world's annual consumption of timber is currently more than 2300 million cubic metres (a well-grown conifer in a European plantation contains around 1–2 m^3 of wood at maturity). This is used for anything from building to paper-making to the creation of chemicals including synthetic rubber. We get a million matchsticks from an average Canadian aspen. Cloth has historically been made from tree bark by the Polynesians and Africans and we can now make rayon from wood. (Incidentally, wooden chopping boards show mild antibacterial properties and so are better than the seemingly more hygienic plastic ones!)

Numerous things we eat and drink come from trees. From the Old World comes citrus fruit, cinnamon (*Cinnamomum verum*), cloves (*Eugenia caryophyllus*), nutmeg (*Myristica fragrans*), coffee (*Coffea arabica* and *C. robusta*), tea (*Camellia sinensis*) and carob (*Ceratonia siliqua*). From the New World we get, among others, papaya (*Carica papaya*), avocado (*Persea americana*), cocoa (*Theobroma cacao*) and the Brazil nut (*Bertholletia excelsa*). There are also oils such as olive oil (*Olea europaea*) and palm oil (*Elaeis guineensis*), the latter from tropical W Africa and used for margarine, candles and soap. The artificial vanilla flavour of cheap ice-cream is a chemical derivative of wood. And cellulose extracted from wood is a common ingredient in instant mashed potato (and disposable nappies). Medicinal compounds from trees are legionary, including quinine (*Cinchona* spp.[1] 'Jesuit's bark' from Peru and Ecuador) used to fight malaria (hence the colonial passion for gin and tonic, a palatable way of taking your bitter quinine medicine). Others are still being discovered: extracts from the ginkgo are currently being advocated for improving blood flow to the brain to improve memory, particularly in those suffering from Alzheimer's disease.

Trees fulfil many other uses. Where would we be without rubber, most of

---

[1] Throughout this book the abbreviation 'spp.' is used for species.

which comes from one species (*Hevea brasiliensis*)? Or brake linings in our cars (which can be made from lignin extracted from wood)? We can even run our vehicles on wood. One tonne of wood can produce 250 l of petrol; thus 4–5 t would keep a family on the road for a year. In urban areas, trees trap and absorb pollutants, and moderate the climate. It is no wonder that studies have shown trees in urban areas to be a contributor to reducing stress and speeding healing in hospitals. A single mature beech tree can produce enough oxygen for 10 people every year and fix 2 kg of carbon dioxide per hour. New uses are still being found for trees. Genes for a sweet-tasting but low-calorie compound called monellin have been taken from a tropical shrub and put into tomatoes and lettuces to make them sweeter.

In the natural world, trees and shrubs often act as nurse plants, aiding the establishment of other species. An example is seen in the Sonoran desert where most of the big saguaro cacti (*Carnegiea gigantea*) start life in the shade of a tree. Animals can also benefit: cows in the Midlands of Britain (and undoubtedly elsewhere) give more milk when they can shelter behind shrubby hedges. In ecological parlance, woody plants are often 'keystone species', those on which many other plants and animals depend.

Parts of the tree

Before we look at different aspects of trees in detail, we should start with an overview of the whole tree.

A tree lives in basically the same way as any other plant. The leaves produce sugars, which are the fuel used to run the tree and used to make the basic building blocks of cellulose and lignin that form the bulk of a tree (Chapter 2). The sugars are moved through the inner part of the bark to where they are needed around the rest of the tree. Sugars not required for immediate use are stored in the wood of the trunk, branches and roots (Chapter 3). The roots at the other end of the tree (Chapter 4) absorb water and minerals (such as nitrogen, phosphorus and potassium) from the soil. The water and dissolved minerals are pulled up through the wood of the tree to reach the leaves, the main users of water. The minerals are used with the sugars to build essential components of the tree, including the flowers and fruits needed to start the next generation (Chapter 5).

Trees get bigger in two ways. The buds scattered around the tree are the growing points for making existing branches longer or for making new branches (referred to as primary growth; Chapter 3). Once made, the woody skeleton gets fatter (secondary growth) by a thin layer of tissue (the cambium) beneath the bark adding new bark and new wood. In temperate areas where growth stops over winter these new layers are seen as the familiar annual rings in the wood (Chapter 3).

The pattern in which new buds are laid down on a developing branch, and which of these buds grow out and by how much, determines the characteristic shape of each tree (Chapter 7). Incidentally, many monocotyledons, such as palms, have one growing tip only and so inevitably live near the tropics where climate is less likely to kill the point: if that growing point dies, so does the whole tree.

The essential difference between plants, including trees, and an animal is that most animals act as a whole unit (one heart, one set of eyes, one liver acting for the whole animal) whereas plants are modular, made up of similar parts added together, each acting largely independently, each replaceable. Thus a tree can lose and replace a branch or even the whole trunk, which for an animal would be equivalent to, for example, cutting me off at the feet and watching them regrow a new me. This modular organisation of trees makes for some interesting problems and solutions in keeping a woody skeleton going for hundreds and sometimes thousands of years. Generally, the living portion of the tree is a thin skin over a long-dead skeleton, which nevertheless must be preserved from the attentions of fungal rot and animals in a number of ingenious ways (Chapter 9).

The story is not yet complete: we can see how a tree is organised within but it must still interact with its environment. It must start from a seed and do battle against a whole army of animals and fungi, and compete with its neighbouring plants. And since trees are long-lived compared to most other plants, they have some neat tricks for surviving (Chapters 6 and 8).

Chapter 2: Leaves: the food producers

Perhaps the most striking thing about tree leaves is their tremendous diversity in size. The Arctic–alpine snow willow (*Salix nivalis*), which grows around the northern hemisphere, can have leaves just 4 mm long on a sprawling 'tree' no more than a centimetre high (Figure 2.1). Smaller still, the scale needles of some cypresses are nearer a millimetre long. Among the largest of leaves are those of the foxglove tree (*Paulownia tomentosa*), which on coppiced trees can be over half a metre in length and width on a stalk another half metre long. Such large sail-like leaves are in great danger of being torn by the wind (as in the traveller's palm, *Ravenala madagascariensis*; see Figure 2.1) so it is perhaps no surprise that big leaves are usually progressively lobed and divided up into leaflets to form a compound leaf. This can lead to even larger leaves: the Japanese angelica tree (*Aralia elata*) can have leaves well over a metre in length (Figure 2.1). Many palms have feathery leaves over 3 m long and in the raffia palm (*Raphia farinifera*) up to 20 m (65 feet) long on a stalk another 4 m long.

The leaves are the main powerhouse of the tree. Combining carbon dioxide from the air with water taken from the soil they photosynthesise, using the sun's energy to produce sugars and oxygen. These sugars (usually exported from the leaf as sucrose, the sugar we buy in packets) are the real food of a tree. They are used as the energy source to run the tree; they form the raw material of starch and cellulose and, combined with minerals taken from the soil, they allow the creation of all other necessary materials from proteins to fats and oils.

The role of leaves in producing food should not be underestimated. A large apple tree holds 50 000–100 000 leaves, a normal birch tree may average 200 000 leaves and a mature oak can have 700 000 leaves. Even this pales somewhat in comparison with the 5 million leaves reported from mature American elms (*Ulmus americana*). Using these vast numbers of leaves a mature beech (*Fagus sylvatica*) can fix 2 kg of carbon dioxide per hour, producing as a by-product enough oxygen for ten people every year. The world's trees have been estimated to produce 65 000–80 000 million tonnes of dry matter per year, two thirds of the total produced by all land plants.

The make-up of a leaf

Broadleaved trees, as the name suggests, have leaves with a broad leaf blade (or lamina) to catch as much light as possible. This limp, fragile material is permeated

(*a*)

(*b*)

Figure 2.1. (a) Arctic willow (*Salix nivalis*) in the Canadian Rocky Mountains; (b) traveller's palm (*Ravenala madagascariensis*) on Tenerife, the Canary Islands, showing the large sail-like leaves that are readily torn by the wind; (c) a leaf of the Japanese angelica tree (*Aralia elata*) near Vladivostok, Russia (the part above the flower-head is all one bipinnate leaf); and (d) welwitschia (*Welwitschia mirabilis*) in the coastal desert of Namibia (photographed by G. Smith).

(*c*)

(*d*)

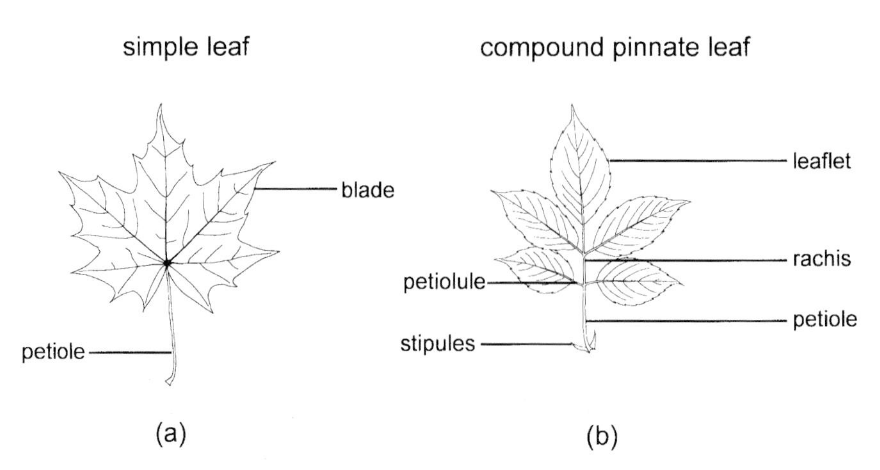

Figure 2.2. Component parts of simple and compound leaves.

by the ramifying skeleton of veins that spread from the leaf base (Figure 2.2). These veins form the bulk of the leaf stalk (which botanists call the petiole). The veins contain strengthening tissue but also the conducting (vascular) tissue, which brings water into the leaf (along the xylem; see Chapter 4 for more details) and takes away the manufactured sugars (through the phloem) (Figure 2.2). Some leaves are accompanied by a pair of small scaly or leaf-like appendages at the base of the leaf stalk: the stipules.

Some trees have simple and entire leaves with no teeth or lobes but others are progressively more deeply lobed, leading to the compound leaf where the blade has been divided up into separate leaflets (Figure 2.2 gives the names of the separate bits of a compound leaf). Trees have three common types of compound leaf: palmate where the leaflets are spread out like fingers from the palm of your hand (e.g. horse chestnut); pinnate where leaflets are attached either side of the stalk (the rachis) which is a continuation of the petiole (e.g. ash, walnut); bipinnate where each leaflet of the pinnate leaf is further divided into leaflets (e.g. acacia and *Aralia* species; Figure 2.1). But what differentiates a compound leaf from a branch with individual leaves? The easy answer is that in broadleaved trees, there is a bud in the junction where a leaf joins the twig (the axil). In a compound leaf the only bud is at the very base; there are no buds in the axils where the leaflets join the rachis.

As is usually the case, however, there are exceptions. In two related tropical genera, *Chisocheton* and *Guarea*, the compound leaves can have flower buds in the leaflet axes and a terminal bud that produces new leaflets. Conversely, and more commonly met in temperate regions, the deciduous dawn redwood (*Metasequoia glyptostroboides*) produces what appear to be pinnate leaves but are actually small simple leaves borne on a short-lived branch that is normally shed at the end of the year. This may explain why the bud that should be in the axil between the 'leaf' and branch is usually to the side or below.

It is sometimes said that a compound leaf can be told from a branch of leaves because the compound leaf falls from the tree as a unit. Anyone looking at an ash tree in autumn will see that this is not necessarily so; leaflets often fall independently.

Inside the leaf

Inside the leaf, the chlorophyll that captures the sun's energy is contained in tiny chloroplasts, which are concentrated in the tightly packed cells of the upper part of the leaf where there is most light. Beneath each square millimetre of leaf there are up to 400 000 chloroplasts, giving a total area exposed to light within a large deciduous tree of more than 350 square kilometres. The leaf needs to be waterproof to maintain a moist working environment inside and prevent the leaf from wilting. To this end leaves have a skin (epidermis) of tightly packed cells overlaid by a waxy cuticle, which has the property of resisting the passage of water without halting completely the movement of air. This is capitalised on in shoe polish, which is usually based on the leaf wax of the Brazilian carnauba wax palm (*Copernicia cerifera*) (and which, incidentally, is also used to coat certain sugar-coated chocolate confectionery!). There is still a need for holes in this waterproof skin to allow effective exchange of gases. In trees, these holes or stomata (stoma, singular) are usually concentrated on the underside of the leaf away from the direct heat of sunlight (especially in glossy, evergreen leaves) but occur on the upper surface (although in low density) in about 20% of tree species (Box 2.1). Stomata can also be found on twigs and branches that have a green photosynthetic layer. The breathing holes of the more mature bark (the lenticels; see Chapter 4) are usually formed beneath old stomata.

Physical laws dictate that most gas can be exchanged for least water loss by having many small widely spaced holes. Because of these constraints, stomata in tree leaves are similar in size and number to those in other plants: typically around 0.01–0.03 mm long and one third that across with densities from as low as 1400 per square centimetre (cm^{-2}) in larch to 25 000 cm^{-2} in apple and over 100 000 cm^{-2} in some oaks, with most leaves falling into the range 4000–30 000 cm^{-2}. Even with such high numbers the stomata still usually cover less than 1% of the leaf area. Usually the stomata are evenly spread over the leaf but this is not always so. In European ash (*Fraxinus excelsior*) stomata are most plentiful in the middle of the underside whereas in horse chestnut (*Aesculus hippocastanum*) they are predominantly around the edge of the broadest parts of the leaflets.

Box 2.1 Examples of trees with stomata on different leaf surfaces

Both sides of the leaf	Eastern cottonwood	*Populus deltoides*
	Black poplar	*P. nigra*
	European larch	*Larix decidua*
Lower side only	Hornbeam	*Carpinus betulus*
	English oak	*Quercus robur*
	Indian bean-tree	*Catalpa bignonioides*
	Oleander	*Nerium oleander*
	Ivy	*Hedera helix*
	Cherry laurel	*Prunus laurocerasus*
Upper side only	Common lime	*Tilia × europaea*
	Blue gum	*Eucalyptus globulus*

Data from Bidwell, R.G.S. (1979). *Plant Physiology* (2nd edn). Macmillan, New York; and Fitter, A.H. & Hay, R.K.M. (1987). *Environmental Physiology of Plants* (2nd edn). Academic Press, London.

Sun and shade leaves

With the large number of leaves held by a tree it is almost inevitable that some will shade others. How the tree gets round this problem is dealt with in Chapter 7. Part of the solution is that the leaves of most trees, even those with open canopies such as birch, can be divided into sun and shade leaves. Shade leaves are larger, thinner, with a thinner cuticle, darker green (more chlorophyll per unit mass), less lobed, with half to a quarter the density of stomata. They can work efficiently at lower light intensities than sun leaves but cannot handle bright light for long periods. However, shade leaves have a quicker response time in opening stomata and so can utilise the brief moving sunflecks that dapple the ground under the canopy. Sun leaves work better in bright light and if put into deep shade their intrinsically higher respiration rates require more carbohydrate than they can produce, so they lose weight and eventually die.

Bear in mind that each leaf develops according to the conditions under which it grows. The proportion of sun and shade leaves thus depends entirely on the growing conditions of the tree. Seedlings may be all shade leaves, acquiring sun leaves only as they emerge above surrounding competition. This can happen within the lifetime of an individual leaf. If a sun leaf is gradually shaded it reacts by 'plastic' changes in shape, becoming thinner and broader. But sudden changes can be more than a leaf can handle. Shade leaves of rainforest seedlings exposed by disturbance or, in Europe, inner leaves of hedges of hazel and hornbeam exposed by clipping can be scorched and damaged by bright exposed conditions.

Water loss by leaves

The loss of water by a tree (called transpiration) can be prodigious. A young apple tree 2 m tall may use 7000 l in a summer; a normal sized birch tree may use 17 000 l and a larger deciduous tree up to 40 000 l. This is equivalent to 40–300 l per day reaching perhaps 500 l per day in a well watered palm or as little as 2–3 l per day in the humid conditions at the bottom of a tropical rain-forest. Trees in dense temperate woodlands (or in 'urban canyons' between tall buildings) use about two thirds the water of a solitary tree. For comparison, at the other end of the spectrum, a large woody cactus may lose only 25 ml (0.025 l) per day. Water loss is, of course, dependent upon surrounding conditions: a well-watered plant in dry warm windy conditions leads to most loss. It is difficult to put trees in order of water requirements but there seems to be general agree-ment that poplars use more than many other trees.

Although water loss is a side effect of having holes in the leaves for gas exchange, it does have its benefits. The flow of water up the tree is the main way of get-ting dissolved minerals taken up by the roots to where they are needed. Moreover, evaporation helps to keep leaves from overheating in hot sunshine. In fact, half the energy of the light reaching the leaves may go in evaporating water. On a larger scale, the vast quantity of water evaporated into the air by forests produce cool moist air which influences rainfall patterns (especially as some trees release prolific quantities of chemicals into the air—think of the overpowering smell of eucalyptus on hot days—which provide nuclei for water droplet and hence cloud formation). On the other side of the coin, trees can little afford reckless water loss. As is explained in more detail in Chapter 4, water is carried up a tree by being pulled from moist areas (the roots) to drier areas (the leaves). This means that to get water, the leaves have to be constantly operating at a low moisture level: in tall trees there is a narrow safety margin between operating and wilting.

Control of water loss

Because most water loss is through the stomata, this is an obvious place to control water movement. Most trees optimise the need for carbon dioxide uptake against needless water loss. Stomata usually close when it is too cold or dark for photosynthesis or when the leaves are in danger of losing too much water and wilting. For example, on a bright day photosynthesis often shows a mid-day dip when sunlight is most intense and the leaves are losing more water than they are supplied, so the stomata close to prevent wilting and damage. A nearby shrub in the shade may happily photosynthesise all day.

Some trees avoid this mid-day dip by keeping the most exposed leaves more or less permanently pointing downwards. (This also helps prevent leaves overheating in the intense light of mid-day.) This can be seen in the tallest tropical forest trees, eucalypts and the desert shrub jojoba (*Simmondsia chinensis*, from which artificial sperm whale oil is produced). The leaves intercept only about half as much light during the hot part of the middle of the day as they do in the relative cool of the morning and evening. This explains why one finds remarkably little shade under a eucalyptus tree from the heat of the mid-day sun. Others, such as the false acacia, fold up their leaflets at the top of the tree if conditions get too tough.

Control of stomata has to be quite complex. For example, evergreens need to prevent opening of their stomata in winter in warm sunny spells before the roots can warm sufficiently to take up water. In the early part of the century two German foresters, Büsgen and Münch, found that if cut twigs of evergreens were placed in water in a warm room in winter, the stomata of holly opened in a few hours, whereas those of yew (*Taxus baccata*) took a week, and those of ivy (*Hedera helix*) and box (*Buxus*) remained closed.

So delicate is the balance between gas movement and water loss that the density of stomata has been found to change in response to external conditions. Measurements from fossil leaves show that 140 000 years ago the dwarf willow (*Salix herbacea*) had 40% more stomata than the 12 500 stomata cm^{-2} found today. Modern olive leaves (*Olea europaea*) have 52 000 stomata cm^{-2} but leaves from Tutankhamun's tomb (1327 BC) had 35% more. In both cases the reduction in density over time appears to be in response to increasing carbon dioxide concentrations—fewer holes are needed to let in the gas. This decrease has also been seen to happen over just a few decades by comparing herbarium specimens from the 1920s with modern-day material. A similar response has been seen over a much shorter time in response to carbon dioxide and temperature. Compared with leaves produced in the comparative coolness of spring, oak leaves growing in a warm late summer have 15% fewer stomata.

Having said all this, trees vary widely in their ability to regulate water loss. Some trees, for example many eucalypts and European alder (*Alnus glutinosa*), cannot regulate water loss to any great extent by closing their stomata. If you want to drain a wet area, plant these!

In addition to stomatal loss, some water is also lost directly through the cuticle of the leaf. In well-watered leaves this 'cuticular transpiration' may be only 5–10% of total water loss, but in water-stressed leaves (where the stomata are closed) or harsh environments it may be significantly higher. Leaves in sunshine and dry air tend to have more wax in the cuticle and plugs of wax over the stomata. Hairs, such as the felted underside of holm oak leaves (*Quercus ilex*), can help by reflecting light and by increasing the thickness of the boundary layer (the stationary layer of air held over the surface of the leaf) making it harder for water to escape. In the olive, a typical Mediterranean plant, a similar function is carried out by semi-transparent umbrella-like scales covering the lower surface of the leaf. The rhododendrons of the high mountains of Asia, facing cold, dry winters and hot, humid summers, nicely illustrate the subtlety of adaptation. The underside of leaves of most species has a more or less dense covering of hairs, scales or small waxy pegs. During winter, this layer helps to prevent water loss, aided in some species by resins and gums secreted from glandular hairs which dry to further seal the underside of the leaf. In the humid summer, the leaf covering *aids* transpiration by keeping the surface of the leaf below the hairs free from drops of water and the stomata unblocked. The glandular hairs secreting resins may suck water from the leaf, helping water loss and the flow of minerals up from the roots. Certainly, many plants, including a few trees, can exude drops of water from the leaves (guttation). This guttation is driven by roots pushing water up the stem (see Chapter 4 for a discussion of this) and so is seen particularly in small trees (under 10 m tall) in humid environments where there is little evaporation. The droplets may appear from the stomata, from special glands at the ends of the leaf veins (hydathodes), or even from lenticels on branches.

Trees living in or near salty water, such as mangroves or tamarisk species, lose water through special salt glands where briny water is excreted as a way of getting rid of excess salt.

Antitranspirants, sprayed on plants by gardeners, are designed to reduce water loss either by causing stomatal closure or by adding an impermeable film over the plant. Treatment of fruit trees may increase fruit size (by making more water available to the fruit) but will generally reduce photosynthesis as well by preventing carbon dioxide getting into the leaf.

In a further attempt to reduce water loss, desert plants such as woody cacti have evolved to take up carbon dioxide at night when evaporation is least and to use this stored carbon dioxide during the day while keeping the stomata firmly shut (called Crassulacean Acid Metabolism or CAM). This is rare in woody plants and found only in some species of the tropical strangler genus *Clusia* in Venezuela, and then only when the individual is short of water. Because the seed germinates in the canopy of a host tree, it can be very short of water until the roots reach the ground.

Another variation is found in plants with C_4 photosynthesis (as opposed to the normal C_3 photosynthesis) where carbon dioxide is very strongly absorbed even at low concentrations so that photosynthesis can proceed at a high rate when the stomata are nearly closed, hence reducing water loss in hot dry environments. This modified way of growing is found in a number of shrubs in dry areas around the world but is rare in tall shrubs and trees, numbering just several species in the Chenopodiaceae and Polygonaceae in the Middle East and an astonishing diversity of euphorbias in Hawaii.

Taking things in through the leaf

In *Rhododendron nuttallii* and other species, dye placed on the scales on the underside of the leaf was seen to quickly pass down the scales and into the leaf. Such leaves may therefore act to absorb moisture. This is seen even more vividly in the welwitschia (*Welwitschia mirabilis*), a strange tree that grows in the coastal desert of Namibia (Figure 2.1). It is really a type of underground conifer with just two leaves, which grow strap-like from the top of the stem, reaching 8 m long with an area of up to 55 square metres. As the leaves grow from the base they die off at the other end. Apart from the occasional downpour, the coastal strip of desert where these plants grow is practically devoid of rain but it has fog on 300 days of the year. Water is absorbed through the stomata into the plant. The stomatal density at 22 200 per square centimetre is fairly average for trees but it does have as many on both sides of the leaf. The waxy cuticle is surprisingly thin for a desert plant but it has crystals of calcium oxalate in the cuticle, which probably play a role in reflecting the sun's energy.

Nutrients can also be taken up through the leaf. Nitrogen compounds in all forms (gaseous, liquid and particulate) have been seen to be taken up through the cuticle or stomata of trees. In Central European forests this absorption of nitrogen through leaves may contribute 30% of the total nitrogen demand of the trees.

Leaf movements

Leaves are on the whole flexible enough to bend and give before the blast of rain and wind. But most woody plants can move their leaves by themselves, partly by growth of the shoot and partly by movements of the leaves themselves.

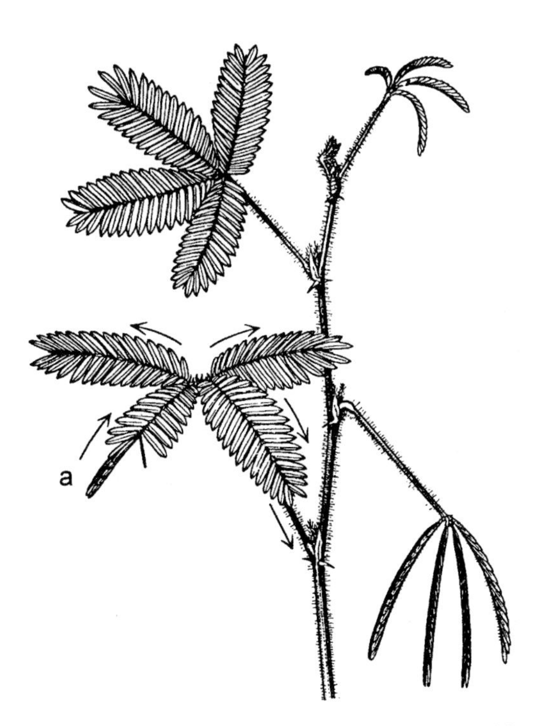

Figure 2.3. The sensitive plant (*Mimosa pudica*) has feathery compound leaves, which can collapse in just 1/10 of a second when touched. If touched at (a) the signal travels in the direction of the arrows. The leaflets fold forward, then the whole leaf droops (as on the right). Movement is controlled by the swollen bases of the leaflets and leaves (the pulvinus). From: Troll, W. (1959). *Allgemeine Botanik*. Ferdinand Enke, Stuttgart.

Perhaps one of the best known examples of leaf movement in trees, and one of the most rapid, is the sensitive plant (*Mimosa pudica* and several related species), a shrubby legume. Touching the plant results in immediate folding and collapse of the leaves and branches (Figure 2.3). The harder the touch the more of the plant collapses. In doing so the leaves are folded away between nasty-looking thorns, and one reason for this collapse may be to move the tender leaves away from browsing animals. Or it may be to dislodge herbivorous insects as they land, or to fold the leaves away from the pounding of rain drops.

The mimosa also shows 'sleep movements' where its leaves are folded down at night. This may have the advantage that when the leaves are not needed for photosynthesis there is value in tucking them away between thorns. Sleep movements can be seen in a wide variety of woody plants from the temperate laburnums, to the tropical tamarind (*Tamarindus indica*), to the desert creosote bush (*Larrea tridentata*). It can involve complex movements. Darwin (1880) noted that in Indian laburnums (*Cassia* species) 'the leaflets which are horizontal during the day not only bend at night vertically downwards with the terminal pair directed considerably backwards, but they also rotate on their own axes, so that their lower surfaces are turned outwards'. The real value of sleep movements in unarmed woody plants is not really known but in herbaceous plants it has been found that folded leaves keep the buds marginally warmer, maybe by only 1 °C, but a significant amount. In herbaceous plants needing short days to flower (see Chapter 5), sleep movements cut down the absorption of light from the moon and stars, preventing the plants from being misled into perceiving the days as long.

Leaf movements during the day allow 'light tracking' where the leaves follow the sun around, resetting during the night to await the rising sun, to maximise the amount of light received by the leaves. Even the needles of conifers will move a little to maximise light catching. While some trees permanently dangle their leaves (see 'control of water loss' above), others fold up their leaves in response to too much light or drought to reduce the amount of light hitting the leaves and hence water loss. For example, the black locust folds the leaflets together like praying hands as the sun gets higher until by mid-day the leaflets are together and edge-on to the sun. In the creosote bush the two leaflets of each leaf open outwards during the day, but will stay closed or almost so if the plant is short of water. The leaves have chlorophyll on both sides, so the vertically closed leaves can still photosynthesise but with a reduced leaf area and reduced water loss. Rhododendrons are well known for drooping and curling leaves in cold weather. Nilsen (1990) suggests that the drooping is to protect leaves from high sunlight intensities in cold temperatures from damaging the leaf's photosynthetic machinery and the curling is to help slow the rate of thawing in the morning, preventing damage during daily rapid rethawing. These examples demonstrate how leaf movements during the day help to optimise the compromise between photosynthesis and conserving water during changing daily conditions.

Some leaf movements are not easy to explain. A famous example is the Asian semaphore or telegraph tree 'whose leaves are in an irregular motion all day long. In this plant of the legume family, sometimes all the leaves move in circles and sometimes leaves on one side of the plant stem move up while those on the other side move down. At times it appears to run wild, with some leaves moving upward, others downward, and yet others moving in circles. It has a jerky movement and occasionally stops as if for a short rest.' (Sandved and Prance 1985).

So how are these movements made? Leaves can simply wilt (as in the drooping of rhododendron leaves in cold weather). Some trees possess the ability to alter the position only while they are growing so they become set in the optimum position. But in most plants, repeated movements take place at bulges at the base of leaves and leaflets called pulvini (Figure 2.3). A pulvinus contains cells that can rapidly move water in and out. By shunting water from cells on one side of the pulvinus to those on the other, rapid movement of the petiole and leaf can be effected. This explains the movement of the individual pulvinus, but how is the message to collapse spread so quickly in mimosa? The mechanism behind this animal-like movement is nerve-like electrical impulses carried not along nerves but through the vascular tissue (see Simons (1992) for details of how this and other plant movements work). The signal can travel at 1–10 cm per second, very impressive for a plant.

Leaf shape

The conflicting needs to intercept light and take up carbon dioxide on the one hand, and to conserve water on the other, have resulted in a number of different leaf shapes in different environments.

A note of caution to begin. So many factors govern leaf shape that it can be difficult to make sense of what we see. Leaf size and shape varies according to the conditions under which they grow, including day length, temperature and moisture, nutrition, and herbivore damage to the leaf and elsewhere in the same or neighbouring trees. Leaves growing later in the summer can be progressively more deeply lobed as in sweetgum (*Liquidambar styraciflua*; Figure 2.4) or longer and narrower as in the lammas growth of oaks (Chapter 6). Add to this that leaf shape changes through a tree canopy and it is perhaps not surprising that it is sometimes very difficult to identify a tree from a leaf. Sometimes there is so much apparently random variation within a single tree or species that shape seems to

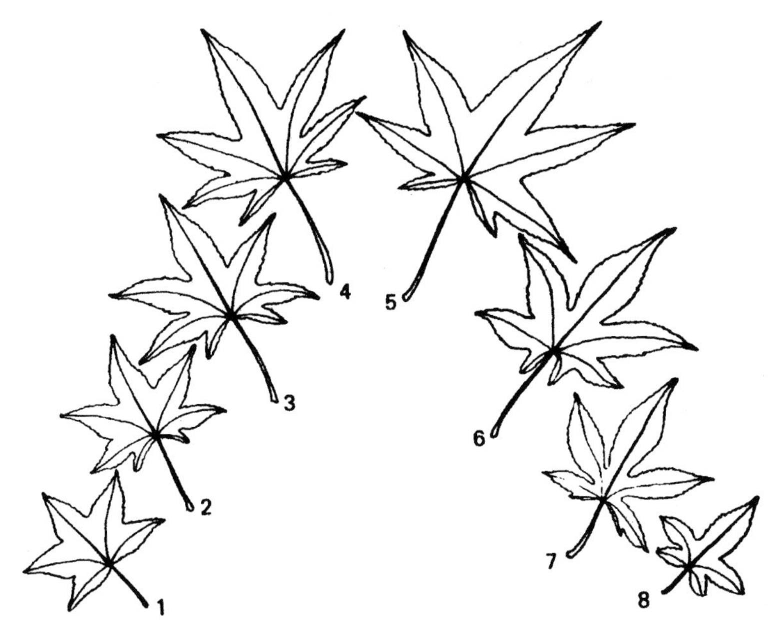

Figure 2.4. The shape of sweetgum leaves (*Liquidambar styraciflua*) from the earliest leaves (1) in spring to the last formed (8) in the late summer. Note the deeper lobing through the season. From: Zimmermann, M.H. and Brown, C.L. (1971). *Trees: Structure and Function.* Springer, Berlin, Figure I-21, Page 49.

have little significance. Despite this wealth of variation due to genetics and growing conditions, a number of generalisations can be made about leaf shape.

Large flat leaves are obviously good at catching light. Moreover, they hold a thick still layer (boundary layer) of air over the leaf, which thermally insulates the leaf leading to a temperature 3–10 °C higher than that of the surrounding air. This increases photosynthesis but also water loss and can lead to leaf-burning in bright light even in the UK. Large leaves tend therefore to be found in shaded areas in wet and humid places where water loss is not so crucial and there is less risk of tearing, exactly the situation in many tropical forests.

With small leaves (or leaflets of compound leaves), the boundary layer is thinner, air moves more easily, the leaf is kept cool by convection rather than evaporation of water. These leaves are therefore more common in areas where water is more precious, such as the less humid conditions of temperate areas. Lobes and teeth on a leaf create turbulence to destroy the boundary layer, making the leaf effectively smaller still; these are common in clearings in tropical forests where high light intensity and lower humidity require efficient cooling, and on larger-leaved temperate trees. This is further accentuated in trees such as aspen (*Populus tremula*) where the petioles are laterally compressed, causing the leaf to shake in the gentlest of breezes, further removing the boundary layer. This is akin to shaking your hands to dry them, which may account for poplars being some of the most profligate users of water and preferring moist alluvial soils.

In drier areas still, such as Mediterranean scrub where rainfall is seasonal, or northern regions where unfrozen water can be in short supply (especially in spring), leaves become smaller still. Small leaves may also be a response to poor or wet soils, which limit root growth and therefore ability to take up water. This is seen in ericaceous plants such as heathers (*Calluna vulgaris* and *Erica* species).

Plants of dry sunny areas also tend to have thick leaves. Thick leaves are less efficient at photosynthesis because the chloroplast layer is thicker and they shade each other and compete for carbon dioxide, but they produce more food without extra transpiration costs. The olive tree seems to have found a partial solution to the self internal shading. The leaves have hard T-shaped stony cells penetrating the leaf like drawing-pins stuck into the surface. These prongs appear to act as miniature optical fibres piping light into the leaf.

Paradoxically, thick leaves are also found in just the opposite conditions: rainforests. Here the tough glossy leaves are designed to reduce the leaching of minerals by the abundant rain sloshing over the leaves. Rain is encouraged to run off by the glossy surface and the elongation of the leaf tip into a 'drip tip'. This prevents water resting on the leaf and leaching minerals, and the growth of light-robbing epiphytes on the surface. As might be anticipated, tall rainforest trees that emerge from the canopy above others have thick leaves but without drip tips: they are dried rapidly by the sun. But you don't have to travel to the tropics to see drip tips; temperate trees in high-rainfall areas, such as limes and birches, also have them.

Needles and scales

Reduction in leaf size with inhospitable conditions can be seen beautifully in conifers, where needles and scales are the commonest type of leaf. Needles usually occur singly but in pines they are in bundles of two, three or five (which fit together to make a cylinder) borne on dwarf branches (as always seems to be the case in plants, there are exceptions; there can be up to eight needles in a bundle, or as in the pinyon pine, *Pinus monophylla*, and occasionally others, single, cylindrical needles). Internally, the needle is no more than a compact version of a simple leaf (Figure 2.5), beautifully compact and adapted to withstand harsh dry conditions (the benefit of evergreen and deciduous needles is discussed below). The chlorophyll-bearing tissue surrounds the central vein, and is in turn surrounded by a thick-walled epidermis and thick waxy cuticle. The relatively few stomata are set in rows (sometimes on both surfaces, more often only on the lower surface) and are sunken into pits (to hold still air over the stomata), often covered in wax to further reduce water loss; the stomata are seen as lines of white waxy dots along the leaf. Needles therefore lose little water when it is short supply. They are also long and thin to shed snow, and contain little sap for freezing.

In most cypresses and in a few others such as the red pines in the Podocarpaceae (*Dacrydium* species), and the Tasmanian cedars (*Athrotaxis* species) and Japanese red cedar (*Cryptomeria japonica*) in the redwood family (Taxodiaceae), the leaves are even more reduced to scales just a few millimetres long which clasp the stem (Figure 2.6). In some, including many junipers, the free tip is elongated to form a needle.

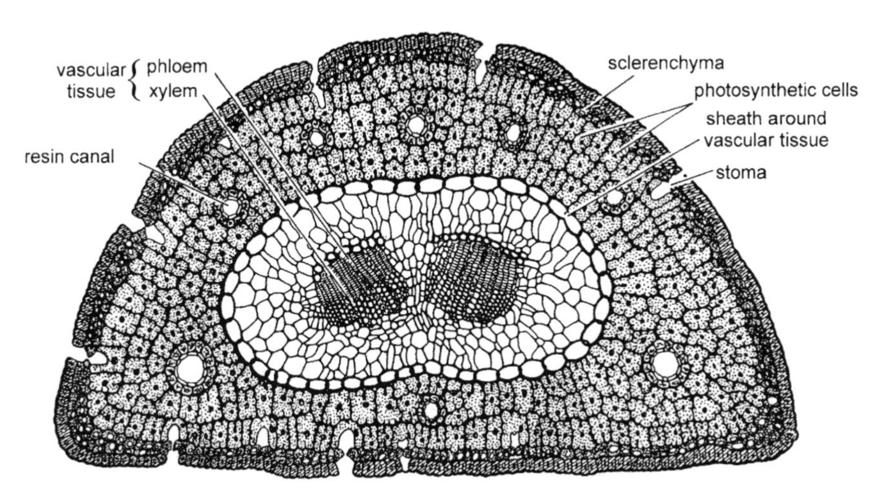

Figure 2.5. Cross-section of a pine needle. From: Foster, A.S. and Gifford, E.M. Jr. (1974). *Comparative Morphology of Vascular Plants* (2nd edn). Freeman, San Francisco.

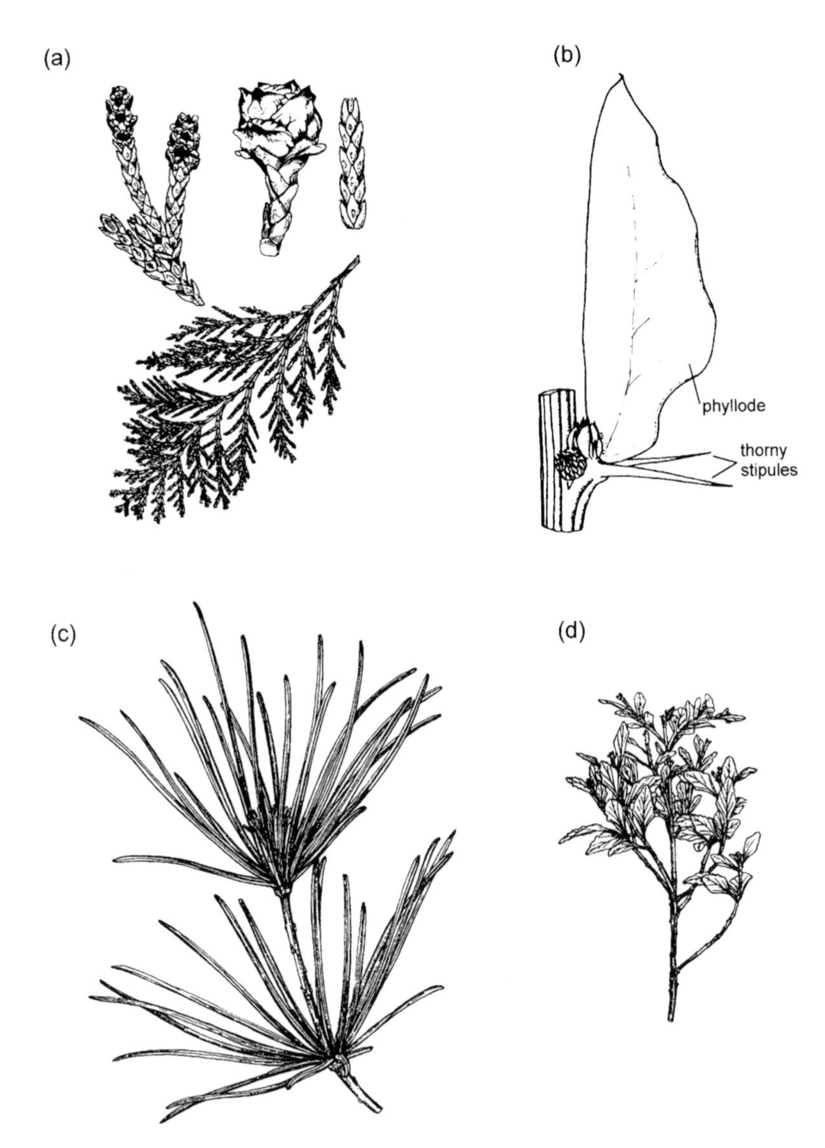

Figure 2.6. (a) Scale leaves of Lawson cypress (*Chamaecyparis lawsoniana*), and (b) a leaf of an acacia (*Acacia paradoxa*) reduced to just a broad flattened petiole (phyllode), above a pair of thorns (formed from the stipules). In (c) the Japanese umbrella pine (*Sciadopitys verticillata*) and (d) the celery-topped pine (*Phyllocladus alpinus*) the leaves are reduced to small brown scales leaving photosynthetic branches. In the celery-topped pine the branches are flattened into phylloclades. From: (a), (c) & (d) Mitchell, A.F. (1975). *Conifers in the British Isles: A Descriptive Handbook.* HMSO, London. (b) Bell, A.D. (1998). *An Illustrated Guide to Flowering Plant Morphology.* Oxford University Press, Oxford. Reprinted by permission of Oxford University Press.

Phyllodes, phylloclades and no leaves at all

Reduction in size of leaves is carried to even greater lengths by some trees. In many acacias, the leaf blade is small or absent and photosynthesis is carried out by broad flattened petioles (phyllodes) (Figure 2.6b). The part of the leaf that loses most water is shed, allowing the plant to continue to function in dry sunny conditions. Some trees and shrubs use leaves when conditions allow but have green photosynthetic stems as a backup during dry periods. This is seen in the creosote bush (*Larrea tridentata*) and other shrubs of North American deserts, and the brooms (*Cytisus* species) of Europe.

Abandonment of leaves as green photosynthetic organs is permanent in some trees and shrubs. In she-oaks (*Casuarina* species), which grow naturally in dry areas of the southern hemisphere, all the leaves are reduced to toothed sheaths surrounding articulations of the stem, producing trees that look like giant, branched horsetails: photosynthesis is carried out by the green cylindrical branches. Similarly, the leaves of the Japanese umbrella pine (*Sciadopitys verticillata*) and the celery-topped pines (*Phyllocladus* species) from Malaysia, are reduced to small brown scales, leaving photosynthetic branches (Figure 2.6c,d). (In *Sciadopitys*, however, there is some doubt as to whether the photosynthetic 'branches' are really branches or two needles joined together!) Branches in the celery-topped pines have taken on the role of leaves in a convincing manner by becoming flattened into green leaflike shoots (called either phylloclades, cladophylls or cladodes, but all meaning essentially the same thing), which look remarkably like pieces of leafy celery. Phylloclades (with and without leaves) are found in a number of other woody plants around the world including the butcher's broom (*Ruscus aculeatus*) in southern England and the Mediterranean (Figure 2.7). It is sometimes possible to find a normal tree, such as a holly, with the occasional fattened branch like a ribbon with leaves. This 'fasciated' shoot is an error in the growth of the branch—an accidental phylloclade—which seems to do little harm.

Many temperate trees and shrubs have chlorophyll in the bark and can photosynthesise during the dormant season. Although this helps offset the respiration cost of the living tissue, and so save on stored food, it does not produce a net increase in food reserves. Desert shrubs on the other hand, which spend most of their year leafless, can produce up to half of their annual food from their green branches.

Other modifications: stipules, tendrils and climbing leaves

In trees that have them, the stipules (Figure 2.2) are often small, scale-like and fall from the tree early in the life of the leaf. In others, such as the Japanese quince (*Chaenomeles* species), they are large and leaflike and add considerably to the photosynthesis of the leaf during its whole life. Yet stipules can fulfil other functions. In the tulip tree (*Liriodendron tulipifera*) and magnolias, the stipules at the base of an expanding leaf stay closed around the young developing leaves at the

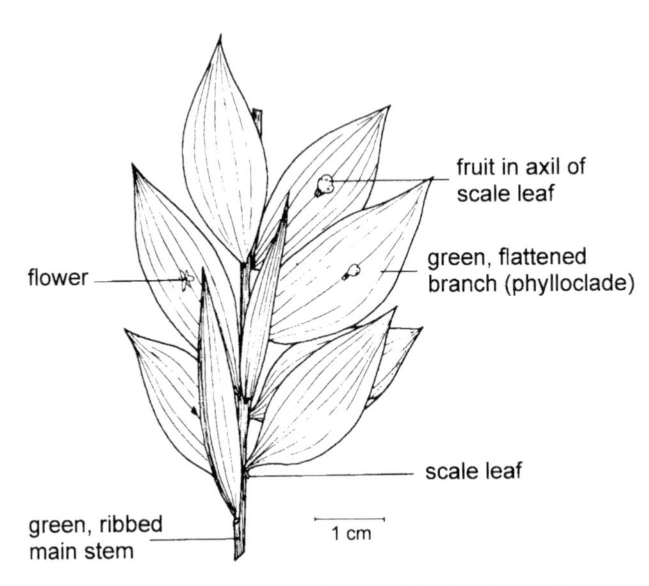

Figure 2.7. The butcher's broom (*Ruscus aculeatus*), which has flattened stems (phylloclades) as the main photosynthetic organs. The leaves are reduced to small scales. This arrangement leads to the flowers appearing to spring from the middle of the 'leaves'. From: Bell, P.R. and Woodcock, C.L.F. (1983). *The Diversity of Green Plants.* Edward Arnold, London.

end of the shoot. Once the next leaf expands the stipules of its protective big brother soon fall. In many trees the stipules form bud scales, e.g. beech, oak, limes, magnolias and many members of the fig family. But in false acacia (*Robinia pseudoacacia*), desert ironwood (*Olneya tesota*) of western United States, and acacias, to mention a few, stipules become woody and persist on the twig as two thorns for defence (see Figure 9.1c).

Leaves can also be used to help woody climbers on their way. In Virginia creeper (*Parthenocissus quinquefolia*) and grapevines (*Vitis* species) the climbing tendrils are modified stems (they grow out from buds). In clematis, as described by Darwin, it is the leaf stalk that twists. The leaves are normal but the sensitive petioles are clasping and wrap around an object. Rattan palms (*Calamus*) of tropical Asia use their leaves in a similar way. The upper leaflets are barbed 'grappling hooks' that catch onto the bark and stems of trees, helping the vine-like palms scramble to the top of the forest.

Juvenile leaves

Changes in shape can also occur as a tree matures. Certain trees go through a number of completely different leaf shapes and arrangements. Leaves of young plants sometimes differ from those of the mature plant in both form and arrangement. For example, many eucalypts, pines and junipers produce juvenile leaves before

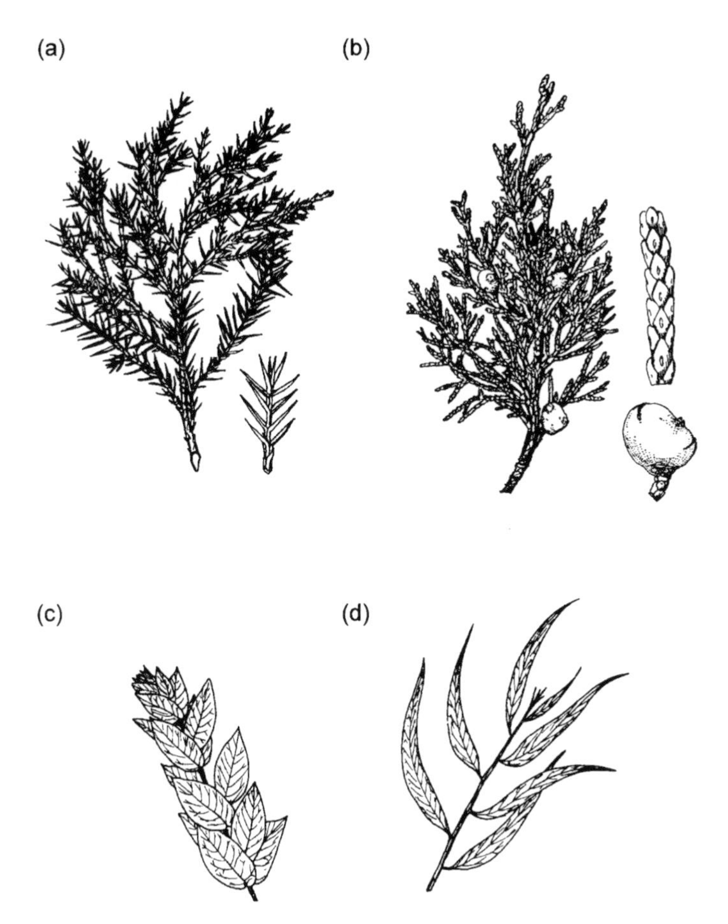

Figure 2.8. Juvenile (a) and adult foliage (b) of Chinese juniper (*Juniperus chinensis*) and blue gum (*Eucalyptus globulus*) (c) and (d), respectively. From: (a), (b) Mitchell, A.F. (1975). *Conifers in the British Isles: A Descriptive Handbook.* HMSO, London. (c), (d) Bell, P.R. and Woodcock, C.L.F. (1983). *The Diversity of Green Plants.* Edward Arnold, London.

the mature foliage (Figure 2.8). In eucalypts the juvenile leaves are round and stem-clasping in comparison with the slender stalked mature leaves. A young tree often has a cone of juvenile leaves inside the canopy, showing that the apices all changed to mature foliage at the same time, usually at around five years of age. In many hardwood trees, juvenile foliage is not a completely different type from mature foliage but a different size and shape. The tulip tree illustrates this nicely. It goes through a juvenile stage up to about ten years of age, during which the leaves can be 25–30 cm long, reducing to 10–15 cm on a mature tree. The shape also changes from four-lobed through to ones with an extra pair or two of lobes (this may be related to sun and shade leaves discussed earlier). The same sequence is restarted with coppicing, sometimes with even greater disparity in size.

Evergreen versus deciduous

The difference between deciduous and evergreen trees is quite clear but holds the seeds for tremendous confusion! Deciduous trees keep their leaves for less than one year and are bare during an unfavourable season (winter in temperate areas, summer in Mediterranean climates). This includes most temperate and some tropical hardwoods, and a number of conifers including dawn redwood, swamp cypress and larches. Evergreen trees by contrast have some leaves all year round. Evergreens include most conifers and also many hardwood species of the tropics, mountains shrubs (e.g. rhododendrons) and temperate trees (e.g. holly). Most conifers and hardwood temperate trees (such as holly) keep leaves for 3–5 years but yew, firs and spruces may keep needles for up to 10 years. Some are more exceptional: both the bristlecone pine (*Pinus aristata*, the oldest trees in the world; see Chapter 9) and the monkey puzzle (*Araucaria araucana*) regularly keep leaves for 15 years and exceptionally for over thirty (the lifespan of a horse). Evergreen leaves usually live for longer (by 2–3 years) at higher altitudes. At the other end of the size spectrum, evergreen shrubs, such as heathers, keep leaves for just 1–4 years (Karlsson 1992). An evergreen canopy is therefore made up of leaves of differing age, perhaps particularly at increasing latitude and altitude, and on poorer soils, where leaves are generally kept longer. In most trees the age of a leaf can be worked out by counting back the number of years of annual growth. It is harder to age leaves in tropical trees with continuous growth but marking experiments have shown that 6–15 months is probably the normal age limit, with some living up to 2–3 years.

Long-lived evergreen leaves do not necessarily stop growing in the first year they are made. Pine needles can continue to grow in length and thickness from the base for many years. Monkey puzzle leaves are attached to the branch along a long base and living for several decades they have to grow wider at the base to remain attached to the expanding branch.

The distinction between deciduous and evergreen habit is not always straightforward. Evergreens such as eucalypts and citrus species are partly summer-deciduous; they may shed some leaves when under the stress of a hot dry summer. As one might expect, it is the older shaded leaves that go in competition with younger more vigorous leaves. In Britain the native privet (*Ligustrum vulgare*) is similar: in mild winters more than half the leaves may survive, but in severe winters they are all shed. (In many evergreen species it is spring drought that causes most problems; see Chapter 9.) Temperate and tropical trees can further flaunt our tidy definitions by being 'leaf-exchanging'. Here new leaves appear around the time that the majority of old leaves fall, either just before or just after (this can vary even within the same species), so some specimens are almost completely leafless for a brief period. This happens in a number of tropical and Mediterranean species including evergreen magnolia (*Magnolia grandiflora*), camphor laurel (*Cinnamomum camphora*), avocado (*Persea americana*) and various oaks including the cork oak (*Quercus suber*). When these leaf-exchanging species are grown in temperate conditions they tend more towards being truly evergreen, but some leaf exchangers can be found under temperate conditions: Spanish oak (*Quercus hispanica*, a hybrid between the deciduous turkey oak, *Q. cerris*, and the evergreen cork oak, *Q. suber*) and Mirbeck's oak (*Q. canariensis*), from Spain and N Africa keep their leaves until early spring, losing them all just before the new set grows. Leaf-exchanging trees could qualify as evergreen because they have leaves for near enough the whole year but also qualify as deciduous because the leaves live around 12 months and change all at once every year. Many young trees in the tropics show continuous evergreen growth but as they get bigger they become leaf-exchanging or deciduous.

Why are some trees evergreen and some deciduous? Starting at the beginning, if there is no unfavourable period for growth during the year, as in the tropics, a plant will usually be evergreen: there is no reason to lose leaves that are still working. In climates with a period unfavourable for growth—a winter or a hot dry summer—deciduous trees have the upper hand. It is cheaper to grow relatively thin, unprotected leaves, which are disposed of during the inhospitable season, than to produce more robust leaves capable of surviving the off season. These seem to be optimum for warm temperate areas (but from the tropics to the Arctic there are always at least a few hardwood deciduous trees and shrubs to be found).

In areas with even worse growing conditions, however, evergreen leaves will reappear. Firstly, this includes areas where the growing season is very short, as in northern and alpine areas. Evergreen leaves start photosynthesising as soon as conditions allow and no time is wasted while a new set of leaves is grown. For the same reason, in European woodlands holly and ivy benefit from being evergreen; they are usually shaded by taller trees and need to do a lot of growing in early spring and late autumn while the trees above are leafless. Secondly, evergreen leaves are found where nutrients are in short supply on poor soils (such as on sandy soils and much of the tropics) or very wet soils where rooting is restricted. Here it is too expensive to grow new leaves each year and it is cheaper to pay the cost of producing more robust leaves that can survive the inhospitable period. Finally, evergreen leaves are found in dry areas such as around the Mediterranean, for example as in the holm and cork oaks. Thick leathery leaves are needed during the fairly dry winter growing season just to survive. This makes them tough enough to withstand the hot dry summer as well. So much energy and nutrients are invested in the leaves that it makes economic sense to hang on to them for several years.

Last of all, if the growing conditions become even more severe, such as in the far north and upper alpine areas, trees may again revert to being deciduous. Despite the problems of having to cope with a short growing season and few nutrients, the winter is so severe that being deciduous is still a cheaper strategy than trying to keep leaves alive over the winter. Thus, the northernmost trees in the Arctic are birches and the final trees up mountains are often birches, willows and larches.

Death and senescence

Deciduous trees shed the bulk of their leaves before an inhospitable period primarily in response to decreasing day length, temperature and soil moisture. Tropical trees are more complicated. Trees may lose their leaves together as a species or as an isolated tree (although there is a tendency for trees within a species to become more synchronised as they become older). Or they may shed leaves from one or more branches at a time, giving the tree a mottled look. The loss may be at irregular times or regular but not necessarily linked to the 12 month calendar. Trees may be leafless for a few weeks to six months. Leaf shedding may be in response to an approaching dry season or even a rainy season (cloudier and darker?).

There is the probably apocryphal story told by Edlin (1976) about the enterprising park keeper who said he would only plant evergreens since they didn't lose their leaves and he would therefore be saved the chore of sweeping. He would have been somewhat disappointed! They may lose only a proportion of their leaves each year but they do shed. In some evergreens leaf fall is spread through the year (with perhaps a midsummer peak as in *Cupressus* species) and so goes largely unnoticed. In contrast others drop their oldest leaves in one go, such as holly in spring and early summer.

In evergreen and 'leaf-exchanging' species it is probably not day length but competition from new leaf growth that stimulates senescence of oldest leaves. This is an important point: leaves act as independent units, similar to a block of apartments. If a tenant is not paying their rent, they are thrown out; if a leaf is a net drain on the tree—it is using more energy than it produces—it is shed (see Chapter 7).

Leaf shedding, for whatever reason, is not just a case of leaf death: if a branch is snapped or the leaves are killed by sudden stress, the leaves wither in place but are remarkably hard to pull off. Leaf fall involves a carefully executed severance similar to the way other parts of trees such as fruits, flower parts and even branches are shed. Across the base of the leaf and leaflet stalk there runs a line of weakness—the abscission zone—made up usually of small cells lacking lignin (Figure 2.9). At leaf fall the 'glue' holding these cells together weakens until finally the leaf is held on by just the cuticle and the plumbing tubes (vascular tissue). At this point the leaf is usually torn off by wind or forced off by frost. With the latter, expanding ice crystals force the leaf off which drops when the ice melts: this explains the sudden deluge of leaves on a warm autumn morning. As the leaf falls, or even before, the abscission zone seals off the broken end of the branch by producing a corky layer.

Certain trees, including beech, oaks, hornbeams, American hop hornbeam (*Ostrya virginiana*) and sugar maple (*Acer saccharum*), are renowned for holding on to some of the dead withered leaves. Leaves are normally kept on the bottom 2 m of the tree but occasionally up to 10 m. The blade and most of the peti-

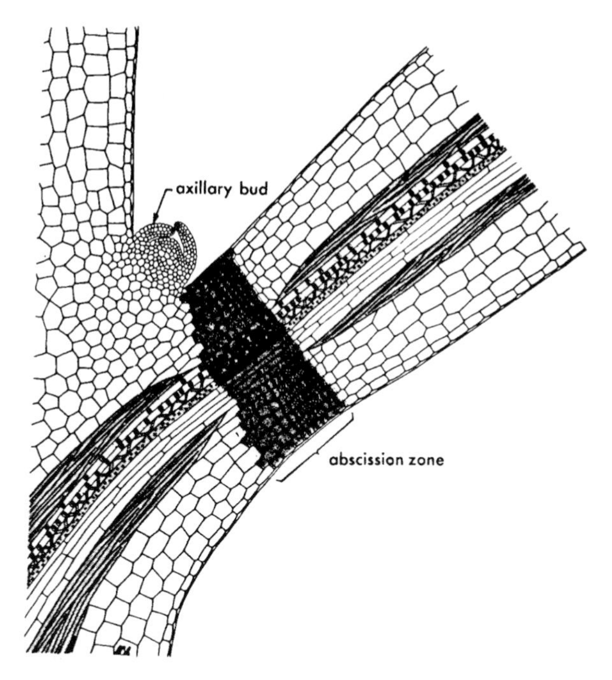

Figure 2.9. Abscission zone made up of corky cells at the base of a leaf. From: Weier, T.E., Stocking, C.R., Barbour, M.G. and Rost, T.L. (1982). *Botany: An Introduction to Plant Biology* (6th edn). Wiley, New York. Reprinted by permission of John Wiley & Sons, Inc.

ole die but the very base remains alive until the following spring, when abscission is completed and the leaf falls. Wind may rip away some leaves in the winter but the petiole base doggedly hangs on. It seems that these trees are rather slow at beginning abscission in the autumn and it gets halted half way through by cold temperatures, only to resume where it left off in the spring. But why? It may act to keep the buds of young growth protected from frosts. But Otto and Nilsson (1981) found that the spring-shed leaves of beech and oak were richer in soluble minerals than autumn-shed leaves that had been lying on the ground over winter, presumably because leaves held on the tree were drier and therefore less readily attacked by fungi and microbes. When the leaves fall in spring the roots are active and able to take in these nutrients as they leach from the leaves. Keeping the leaves only on the lower branches improves the chances that they will fall within reach of the tree's own roots! Retention of dead leaf bases is also a normal feature of palms and tree ferns.

Leaf colours

Most leaves look green because photosynthesis is driven by blue and red light and it is the green that is left over to be reflected or passed through the leaf. But not all leaves are green. New leaves, especially in shade-tolerant tropical trees but also the majority of temperate trees, can be very pale green to white, various pinks and brilliant reds in maples, to brown in cherries, beech, southern beeches, poplars, the katsura, tree of heaven and *Catalpa* spp.). Here other pigments (particularly anthocyanins, which also give the rich colours of *Begonia rex* and *Coleus*), are formed before the chlorophyll. The exact significance is difficult to identify but Kursar and Coley (1992) suggest from work in Panama that because tender young leaves are prone to being eaten (they typically receive 60–80% of their lifetime herbivory when young), the trees do not put valuable chlorophyll into leaves until fully expanded and tough.

In naturally occurring variants of some trees, such as the copper beech and sycamore, anthocyanins occur sufficiently intensely to mask the chlorophyll for the life of the leaf. Anthocyanin production requires light, so the reddest leaves are on the outside of the canopy and the inner shaded ones can be almost a normal green. This masking of the chlorophyll does not appear to be much of a handicap.

Pigments are used in other ways. Leaves below flowers (officially called bracts) are modified in dogwoods and the dove (handkerchief) tree (*Davidia involucrata*) to a brilliant white to help attract insects to the diminutive flowers. Variegated leaves are much prized by gardeners. The cause may be genetic mutations (including chimaeras; see Chapter 8), viral (as in tulips), or mineral deficiencies.

Perhaps the most familiar colouring of leaves is with the autumn. Some fall green (e.g. alder and ash) but many produce a mixture of bright yellow, orange and red. In eastern Canada and New England the autumn colours are spectacular. Yet the same species grown in Europe rarely attain the same intensity. Why?

Some of the blame can be put down to genetic differences between individuals. A row of sibling seedlings can vary tremendously in the intensity and duration of autumn colour. We may be just unlucky with our choice of seedlings. But autumn colours are so reliably less good in Europe no matter what we plant that there must be something else.

As a leaf starts to die in the autumn, substances that can be reused, such as proteins and chlorophyll, are broken down and taken back into the tree. At the same time as leaves are systematically stripped of useful assets they are filled with unwanted things such as silicon, chlorine and heavy metals. (It has been suggested that leaves are used as a dumping ground for waste products like filling rockets with rubbish and firing them off into space: 'excretophores'). The yellow pigments (carotenoids, i.e. carotenes and xanthophylls) normally found in a leaf but previously masked by the chlorophyll, now show through (a similar process happens in the ripening of bananas and citrus fruit). Red pigments (anthocyanins) mixed with the yellow in different proportions produce the oranges, reds, purples and sometimes blues. These red anthocyanins are commonly produced in great quantities in dying tissue from sugars that remain in the leaf. However, this requires warmth and bright light during the day for the remaining chlorophyll to work, and cold nights to slow the transport of sugar out of the leaf. These are common features of the sunny autumns of eastern N America but less so in the often cool and overcast British autumns. The brown colours so typical of the grey shores of Britain are due to other pigments (proanthrocyanins), which also give the dark colour to heartwood and the pink tinge to some sapwoods.

In Japan, Koike (1990) has pointed out that on trees that invade open areas and produce leaves continually through the year (such as birches; see Chapter 6 for more details) autumn colours begin in the inner part of the canopy and move outwards. Conversely in trees that produce all the year's leaves in one flush in spring (maples, for example), autumn colours start on the outside of the canopy.

Why have compound leaves?

Imagine two trees, one of which has branches with 'ordinary' leaves and the other has compound leaves the same overall size and shape as the first tree's branches. What is the advantage of the one over the other? Givnish (1978) has suggested that the advantage of compound leaves is that they are cheap disposable branches, which give the tree an advantage under two conditions.

The first is that they help reduce water loss in arid seasons. Small twigs (with a large surface area to volume ratio and thin bark) are the biggest source of water loss once leaves have fallen. So compound leaves are common in areas with severe dry summers such as savannas, thorn forests and warm deserts, e.g. the acacias and baobab tree (*Adansonia digitata*) of African savannas, the mesquites (*Prosopis* spp.) and palo-verdes (*Cercidium* spp.) of N American desert grasslands, and horse chestnuts (*Aesculus* spp.). One might also include the ash trees that grow on dry limestone screes in central England. Such water loss is unlikely to be a significant factor in most temperate trees and indeed there are fewer compound-leaved trees.

The other advantage of having cheap disposable branches is seen where rapid vertical growth is important. It is assumed that because the rachis of a compound leaf is held up primarily by water turgor and fibrous material it is not as costly to build as a woody branch. A tree that is growing rapidly can display leaves on cheap branches, which will only be needed for a short time before they too are shaded, and invest the saved energy in height growth. The tree thus avoids making an unnecessarily durable and expensive investment. Compound leaves are seen in many early-successional species that need to grow tall quickly to stay ahead of the competition, especially in tropical trees that invade gaps. Some trees in the family of Meliaceae grow for months or years without branches, using large compound leaves as branches. In temperate regions, sumacs (*Rhus* spp.), tree-of-heaven (*Ailanthus altissima*), Kentucky coffee-tree (*Gymnocladus dioica*) and Hercules' club (*Aralia spinosa*) also fill this niche and all bear few main branches and numerous compound leaves as throw-away 'branches'. Note that others such as box elder (*Acer negundo*) and ashes (*Fraxinus* spp.) are suggested to have compound leaves because they are predominantly floodplain trees and are therefore disturbed area specialists that need to cope with periodic damage.

These proposals are obviously not the whole answer because there are examples of evergreen compound-leaved trees in the moist tropics, and early-successional species with sparse branches, capable of rapid growth, that have large simple leaves, e.g. catalpas (*Catalpa* spp.) and the foxglove tree (*Paulownia tomentosa*). Non-compound-leaved trees have found their own answers. Rapidly growing species such as poplars that have simple leaves do actually shed twigs very readily (Chapter 6); these twigs are thin and not very woody.

Further reading

Beerling, D.J. and Chaloner, W.G. (1993). Stomatal density responses of Egyptian *Olea europaea* L. leaves to CO_2 change since 1327 BC. *Annals of Botany*, **71**, 431–5.

Beerling, D.J., Chaloner, W.G., Huntley, B., Pearson, J.A. and Tooley, M.J. (1993). Stomatal density responds to the glacial cycle of environmental change. *Proceedings of the Royal Society of London* (B) **251**, 133–8.

Brodribb, T. and Hill, R.S. (1993). A physiological comparison of leaves and phyllodes in *Acacia melanoxylon*. *Australian Journal of Botany*, **41**, 293–305.

Bünning, E. and Moser, I. (1969). Interference of moonlight with the photoperiodic measurement of time by plants, and their adaptive reaction. *Proceedings of the National Academy of Sciences, USA*, **62**, 1018–22.

Chabot, B.F. and Hicks, D.J. (1982). The ecology of leaf life spans. *Annual Review of Ecology and Systematics*, **13**, 229–59.

Darwin, C. (1880). *The Power of Movement in Plants*. 2nd edition. J. Murray, London.

Edlin, H.L. (1976). *The Natural History of Trees*. Weidenfeld & Nicolson, London.

Eschrich, W., Fromm, J. and Essiamah, S. (1988). Mineral partitioning in the phloem during autumn senescence of beech leaves. *Trees*, **2**, 73–83.

Ezcurra, E., Arizaga, S., Valverde, P.L., Mourelle, C. and Flores-Martínez, A. (1992). Foliole movement and canopy architecture of *Larrea tridentata* (DC.) Cov. in Mexican deserts. *Oecologia*, **92**, 83–9.

Fisher, J.B. (1992). Grafting and rooting of leaves of *Guarea* (Meliaceae): experimental studies on leaf autonomy. *American Journal of Botany*, **79**(2), 155–65.

Fisher, J.B. and Rutishauser, R. (1990). Leaves and epiphyllous shoots in *Chisocheton* (Meliaceae): a continuum of woody leaf and and stem axes. *Canadian Journal of Botany*, **68**, 2316–28.

Ford, B.J. (1986). Even plants excrete. *Nature*, **323**, 763.

Fromm, J., Essiamah, S. and Eschrich, W. (1987). Displacement of frequently occuring heavy metals in autumn leaves of beech (*Fagus sylvatica*). *Trees*, **1**, 164–71.

Givnish, T.J. (1978). On the adaptive significance of compound leaves with particular reference to tropical trees. In: *Tropical Trees as Living Systems* (ed. P.B. Tomlinson and M.H. Zimmermann), pp. 351–79. Cambridge University Press, New York.

Goldberg, D.E. (1982). The distribution of evergreen and deciduous trees relative to soil type: an example from the Sierra Madre, Mexico, and a general model. *Ecology*, **63**, 942–51.

Karabourniotis, G., Papastergiou, N., Kabanopoulou, E. and Fasseas, C. (1994). Foliar sclereids of *Olea europaea* may function as optical fibres. *Canadian Journal of Botany*, **72**, 330–6.

Karlsson, P.S. (1992). Leaf longevity in evergreen shrubs: variation within and among European species. *Oecologia*, **91**, 346–9.

Kikuzawa, K. (1995). The basis for variation in leaf longevity of plants. *Vegetatio*, **121**, 89–100.

Koike, T. (1990). Autumn coloring, photosynthetic performance and leaf development of deciduous broad-leaved trees in relation to forest succession. *Tree Physiology*, **7**, 21–32.

Kursar, T.A. and Coley, P.D. (1992). Delayed greening in tropical leaves: an antiherbivore defense? *Biotropica*, **24**(2b), 256–62.

Lovelock, C.E., Jebb, M. and Osmond, C.B. (1994). Photoinhibition and recovery in tropical plant species: response to disturbance. *Oecologia*, **97**, 297–307.

Nilsen, E.T. (1990). Why do rhododendron leaves curl? *Arnoldia*, **50**(1), 30–5.

Otto, C. and Nilsson, L.M. (1981). Why do beech and oak trees retain leaves until spring? *Oikos*, **37**, 387–90.

Pearcy, P.W. and Troughton, J. (1975). C_4 photosynthesis in tree form *Euphorbia* species from Hawaiian rainforest sites. *Plant Physiology*, **55**, 1054–6.

Sandved, K.B. and Prance, G.T. (1985). *Leaves.* Crown Publishers, New York.

Shashar, N., Cronin, T.W., Wolff, L.B. and Condon, M.A. (1988). The polarization of light in a tropical rain forest. *Biotropica*, **30**, 275–85.

Simons, P. (1992). *The Action Plant.* Blackwell, Oxford.

Winter, K. (1981). C_4 plants of high biomass in arid regions of Asia – occurrence of C_4 photosynthesis in Chenopodiaceae and Polygonaceae for the Middle East and USSR. *Oecologia*, **48**, 100–6.

Woods, D.B. and Turner, N.C. (1971). Stomatal response to changing light by four tree species of varying shade tolerance. *New Phytologist*, **70**, 77–84.

Zotz, G., Tyree, M.T. and Cochard, H. (1994). Hydraulic architecture, water relations and vulnerability to cavitation of *Clusia uvitana* Pittier: a C_3-CAM tropical hemiepiphyte. *New Phytologist*, **127**, 287–95.

Chapter 3: Trunk and branches: more than a connecting drainpipe

The woody skeleton

What makes a tree different from other plants is the trunk (or bole) and branches making up the woody skeleton. The main job of this tough, long-lasting skeleton is to display the leaves up high above other lesser plants in the battle for light. As well as support, though, the trunk and branches have two other important jobs: getting water from the roots to the leaves and moving food around the tree to keep all parts, including the roots, alive. But is the trunk just a large connecting drainpipe that keeps the two ends of the trees apart? In many senses, yes, but its structure allows it to do many other things that no mere drainpipe could do.

Starting from the outside is the outer bark, a waterproof layer, over the inner bark or phloem (Figure 3.1). The phloem is made up of living tissue that transports the sugary sap from the leaves to the rest of the tree. Inside the bark is the cambium, which, as will be shown, is responsible for the tree getting fatter. Inside this again is the wood proper or xylem. Although seemingly 'solid wood' it is the part of the tree responsible for carrying water from the roots to the rest of the tree. The water moves upwards through dead empty cells. But wood is not entirely dead. Running from the centre of the tree are rays of living tissue (made up of thin walled 'parenchyma' cells), which reach out into the bark (and in some trees there are lines of these living cells running up through the wood as well). As will be seen later these living cells are involved with movement and storage of food and the creation of heartwood, the dense central core of wood that (reputedly) supports the tree as it becomes larger. At the very centre of the tree some trees, but not all, have a core of pith (the strengthening tissue when the shoot is very young and soft).

How the trunk grows

New branches grow longer, as in all plants, from the growing point at the end (the apical meristem). Inside, the central pith and outlying cortex make up the bulk of the new growth. Arranged around the pith are strands of 'plumbing' (the vascular tissue), which have the phloem on the outside and xylem inside (Figure 3.2). Herbaceous plants normally develop no further than this 'primary' growth but woody plants specialise in not just getting longer but also getting fatter. This

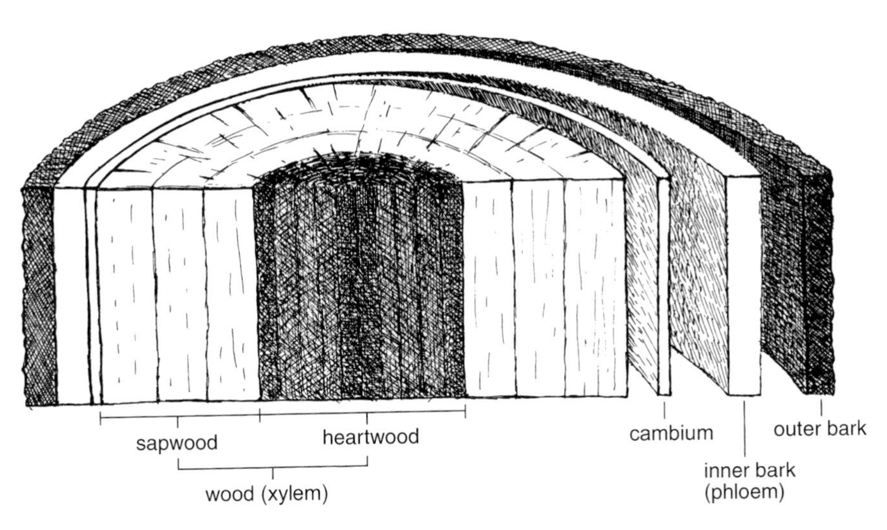

Figure 3.1. Tree cross-section.

is called 'secondary' growth: it happens after the twig has got longer. A branch gets *longer* by adding new primary growth on the end and gets *fatter* by secondary growth. (I finished my primary growth some years ago but I now seem to be undergoing secondary growth, as over the years, my waist expands.) In the young twig, a tissue capable of producing new cells, the cambium (Figure 3.1), forms first between the phloem and xylem and then grows sideways to join up and form a continuous band of cambium around the whole of the inside of the twig. This is joined to the cambium of previous years' growth, forming a continuous sheath over the entire tree under the bark. In theory the cambium is just one cell thick, resembling the gossamer-like sheets of tissue found between the shells of an onion, but in practice it can be a dozen cells or so thick. This fragile, thin, translucent sheet is all that produces new 'secondary' phloem and xylem (and so strictly speaking should be called the vascular cambium: it produces vascular tissue) and is responsible for small seedlings growing in girth to giants of the forest. New xylem (wood) is added onto the inside of the cambium and new phloem on the outside. You will see that as the wood accumulates and the tree grows in girth, it stretches the cambium and the bark. The cambium copes by growing sideways. For the bark (including the phloem) things are a little more complicated (see Bark, below).

Several points are worth making. Note that the oldest wood is in the middle of the tree and gets younger towards the outside; that's why the pith at the centre of a young twig can be found at the centre of a trunk many metres across. In effect as a tree grows it lays down a new shell of wood over the whole tree and the old tree is left fossilised inside. It is possible to dissect a tree and see how it used to be through its life (of course, broken branches will have been lost so

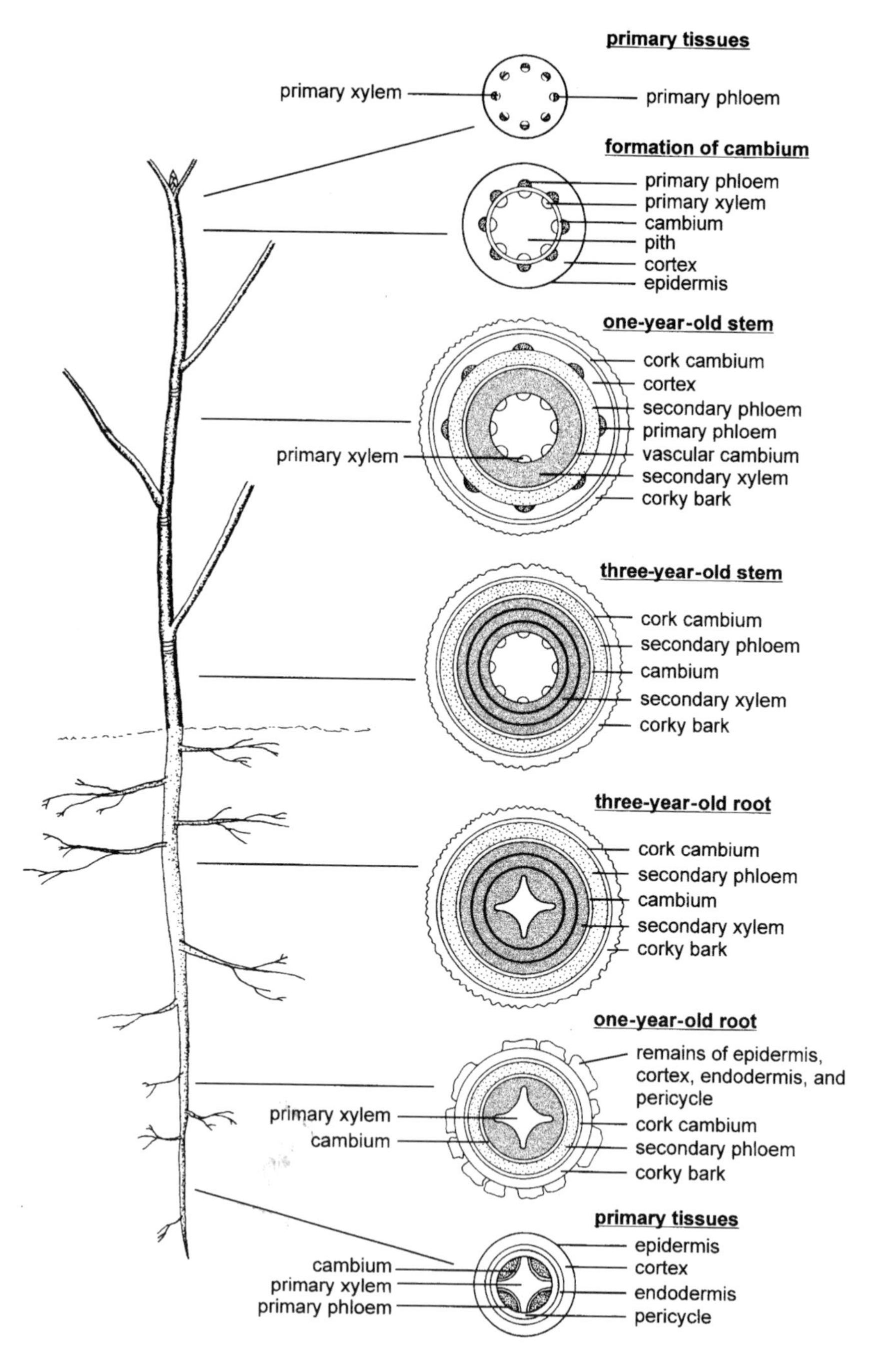

Figure 3.2. Development of a young tree showing how the young shoots and roots develop the woody structure (secondary thickening).

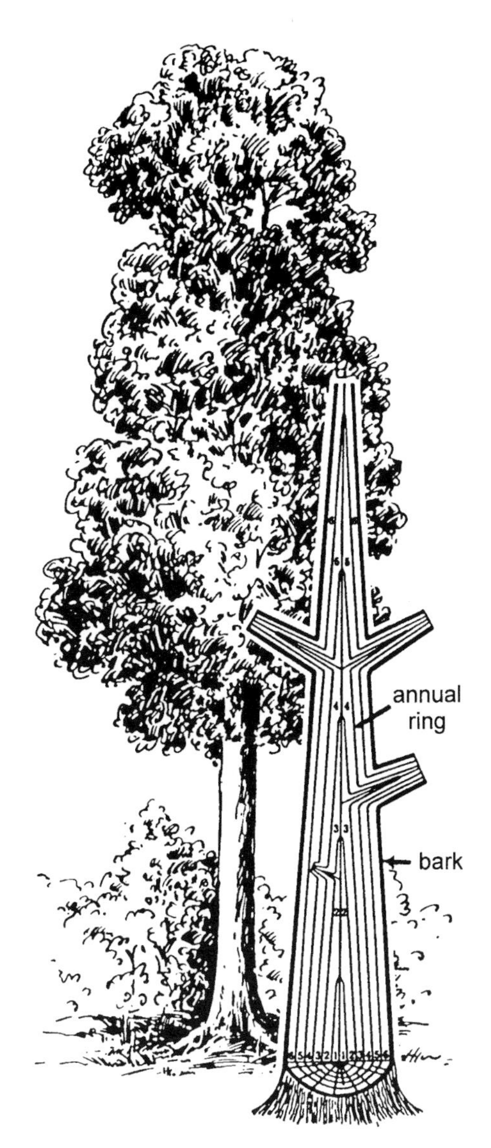

Figure 3.3. A vertical section through a tree showing that it grows by adding on new shells of wood over the old tree. From: Kozlowski, T.T. (1971). *Growth and Development of Trees:* Vol. I, *Seed Germination, Ontogeny, and Shoot Growth.* Academic Press, New York.

the complete tree may not still be there; see Figure 3.3). There's an important corollary to this: trees do not grow by stretching upwards, they put on new layers over the old. If I stab a knife into a tree and come back in ten years' time, the tree will have got fatter (and may be burying the knife) but since the sharp end is embedded into an old part of the tree, which is the same size as it ever was, the knife will still be at the same height above ground. If you find this hard to believe, try stretching a plank of wood. If you can't do it, neither can the tree! Stories like the proverbial 'my swing tied to a branch is further away from the ground than when I was a child' can be the result of the ground getting lower by erosion. But also, branches, being sideways, will lift things further away from the ground as they get fatter.

Not all trunks grow as perfect cylinders (or cones); for example, hawthorns (*Crataegus* spp.) and yews (*Taxus* spp.) tend to have very fluted trunks, resembling a series of trees fused together. In these trees some parts of the cambium are much more active, growing wood more quickly, leading to that part of the tree bulging out. This can become so extreme that two bulges meet up, press together and fuse, leaving a patch of bark included in the wood. The reason for fluting is largely unknown but does appear to be often genetically determined. In lianas, where the biggest problem is not supporting the weight of the foliage but resisting the twisting pressure, fluting is very common. Lianas often have flat ribbonlike or star-shaped stems caused by the relative inactivity of parts of the cambium, allowing them to twist and turn.

Do all trees get wider?

The general answer is that almost all do, one way or another. Tree ferns, which you might argue are not really trees, are a major exception and don't. Monocot plants usually have needle-like bundles of vascular tissue scattered across the stem (hence palm wood is often called porcupine wood). Some, like palms and screw pines (*Pandanus* spp.), do not get fatter by secondary thickening as above but they may get fatter at the base by growing extra parenchyma cells (but without any new vascular tissue) or by hiding their skinny nature under layers of old leaf bases or masses of adventitious roots (see next chapter). Other monocots (see Box 1.1), such as the Joshua tree (*Yucca brevifolia*), dragon trees (*Dracaena* spp.), century plants (*Agave* spp.), cabbage palms (*Cordyline* spp.) and the grass trees of Australia (*Xanthorrhoea* spp.), are capable of secondary growth of a sort. Here, new needles of vascular tissue arise vertically through the stem, making the stem fatter, like pushing pencils down into a cylinder of foam rubber. The palm-like cycads (really primitive conifers) do have proper secondary growth from a succession of cambia and develop loose-textured growth rings around a large pith, but it is a very slow process.

Incidentally, a tree fern or palm 20 cm thick cannot start this thick as a seedling, so how do they get to be this size without secondary thickening? As the seedling grows, successive bits of stem between leaves (the internodes) are slightly wider than those below: the growing point grows simultaneously in width and length. As these internodes are relatively short the mature stem diameter is quickly reached near the base and stability is maintained by adventitious roots.

What is wood made of? Wood structure

Cellulose is the building material of all plants; lignin is the material that makes plants woody. The main constituents of wood are: 40–55% cellulose (the main constituent of paper), 25–40% hemicelluloses (a very diverse group of cellulose-like compounds) and 18–35% lignin. Conifers tend to have more lignin and less cellulose and hemicellulose than hardwoods.

As explained in Chapter 1, there are many differences between conifers (by which is really meant the gymnosperms) and the hardwoods (angiosperms). One profound difference is in the structure of the wood.

Conifers

As can be seen from Figure 3.4, the wood of conifers is composed chiefly of dead empty tubes called tracheids that fit tightly together. These tracheids are the tubes through which water flows up, but since they make up 90–94% of the wood volume they also hold the tree up. The other 6–10% of the wood is made up of rays (these are sheets of living cells, one cell wide in gymnosperms, that run along the radii of the wood).

If you look at a polished piece of pine, it is hard to believe that there are tubes and that it is not just 'solid wood' (any obvious holes seen, either running parallel to the tracheids or through the rays, are probably resin canals: see below). This demonstrates that the tubes are very small. In North American woods the diameter of these tubes ranges from 0.025 mm in Pacific yew (*Taxus brevifolia*) up to 0.080 mm in coastal redwood (*Sequoia sempervirens*): that's 40 to 12.5 tubes to the millimetre, respectively). Nor are they very long, varying from 1.5 to 5 mm long on average, reaching up to 11 mm in the monkey puzzle tree (*Araucaria araucana*). Indeed a cubic inch of Douglas fir wood (*Pseudotsuga menziesii*) can contain 3 million tracheids! Thus, in conifers water has to pass through several tens if not hundreds of thousands of tracheids on its journey from the roots to leaf. This is an even more amazing feat when you consider that the individual tracheids have closed ends (as can be seen in Figure 3.4). The trick is that they are joined together by a variety of 'pits'. The commonest type are bordered pits where there is a raised doughnut around a central hole. Inside, a thickened lump (the torus) is suspended by a web of cellulose strands (called the margo) just like a trampoline suspended by a web of elastic cords. These pits act as valves. Water can flow through freely, but if damage to a tracheid lets in air on one side, the pressure difference between the two sides pulls the torus over and blocks the hole. The value of this to the tree is discussed below but it can cause us humans a problem. Timber from trees like spruce, where these sealed (or aspirated) pits are common in dried timber, is particularly difficult to treat with preservative solutions unless the wood is treated green.

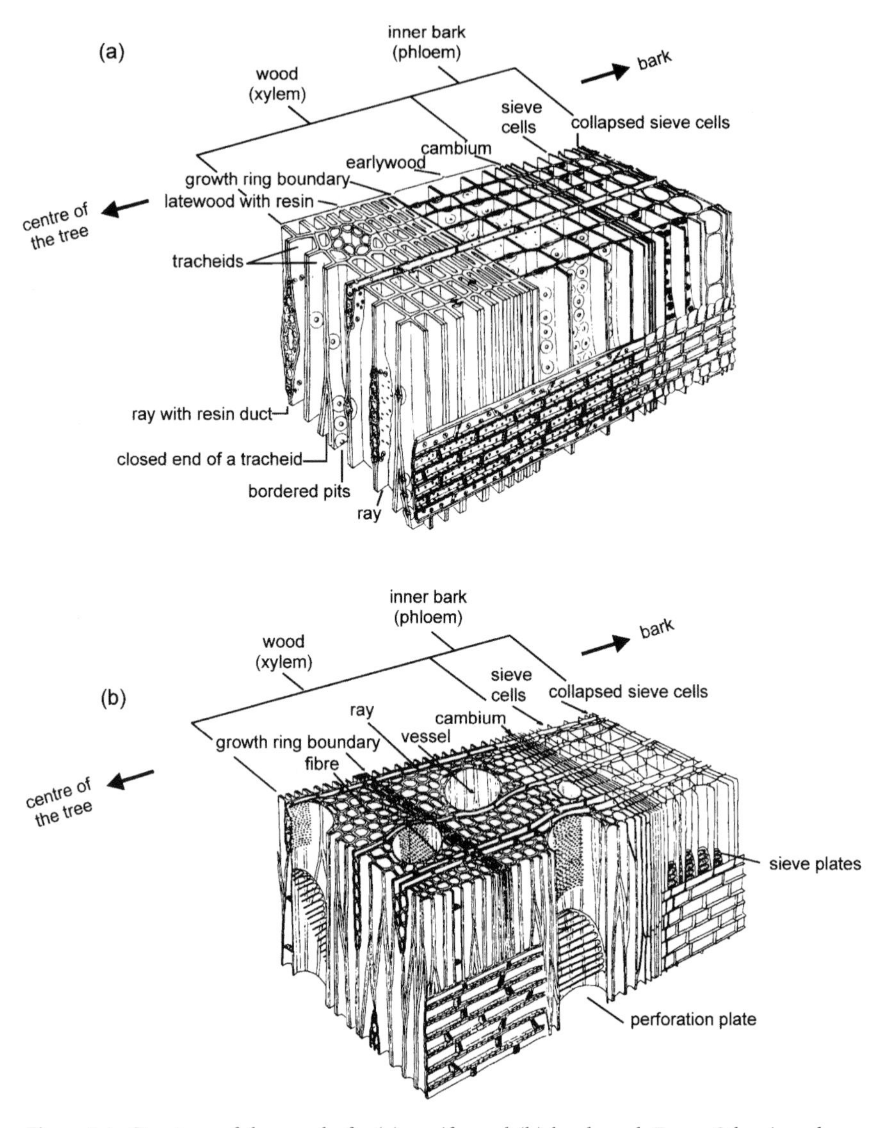

Figure 3.4. Structure of the wood of a (a) conifer and (b) hardwood. From: Schweingruber, F.H. (1996). *Tree Rings and Environment Dendroecology*. Haupt, Berne, Switzerland.

Hardwoods

The wood of angiosperms is more complex, being made up of vessels that conduct water (6.5–55%, average 30%, of the wood volume), small-diameter fibres that provide strength (27–76%, average 50%) and rays (6.5–30%, average 20%). The large-diameter vessels (up to half a millimetre or more wide) are often easily visible on the cut surface (referred to as pores) as in Figure 3.4. The vessels start life as vertical series of cells (called vessel elements) each no more than a millimetre long. As they die the cross-walls largely disappear to leave longer tubes that might be just a few centimetres long or may run the whole length of the tree. The old cross-walls are left as 'perforation plates', which vary from a single round hole to complex 'multiple perforations' in patterns of slits or holes. Where neighbouring vessels meet each other and rays, they are joined by pits similar to those in conifers.

The fibres are long narrow cells with thick walls, a small cavity and tapering ends, closely bound together to form a matrix holding the vessels. The rays are also more complicated in hardwoods. As in conifers they may be just one cell wide (uniseriate) as in willows and poplars, but are commonly many cells thick (multiseriate) and up to several millimetres wide as in oak. In addition, many hardwoods have strands or sheets of ray material (parenchyma cells) running vertically through the wood between or around the vessels (called axial parenchyma). The rays may carry gum canals filled with gums, resins, latex or other material used in the defence of the tree (Chapter 9). Rays also tend to hold the oils that give woods their characteristic smell and taste.

Crystals are also found quite commonly in hardwoods (and sporadically in the pine family), both in the wood and in the bark. These white crystals, usually calcium oxalate, are formed in the parenchyma cells when excess calcium taken up with water from the soil combines with oxalic acid, a common constituent of cell sap. Tropical woods can also contain silica as small round beads that increase in size from sapwood to heartwood: you can imagine how quickly they blunt saws!

Exceptional wood structure

Nothing ever seems clear-cut in biology. For the record, both *Ephedra* and *Gnetum* species (primitive 'conifers'; see Box 1.1 in Chapter 1) have vessels. To even things up, a few hardwoods have tracheids instead of vessels such as the South American and New Zealand genus *Drimys* (we grow Winter's bark, *Drimys winteri*, in Britain) and several East Asian genera.

Growth rings

A feature of both hardwoods and softwoods is the presence of rings, each one normally marking a temporary halt in growth at an unfavourable time of the year: winter in temperate areas. The rings appear as repetitive bands of light and dark wood. In conifers the reason is easily seen under a microscope. Wood cells grown early in the spring (earlywood: Figure 3.4) are large and thin-walled, just right for carrying the large quantities of water needed during the active growth in spring and early summer. As the year progresses, the cells become progressively smaller with thicker walls; the tree needs less water once it has finished putting out new growth and the emphasis changes to producing strong wood.

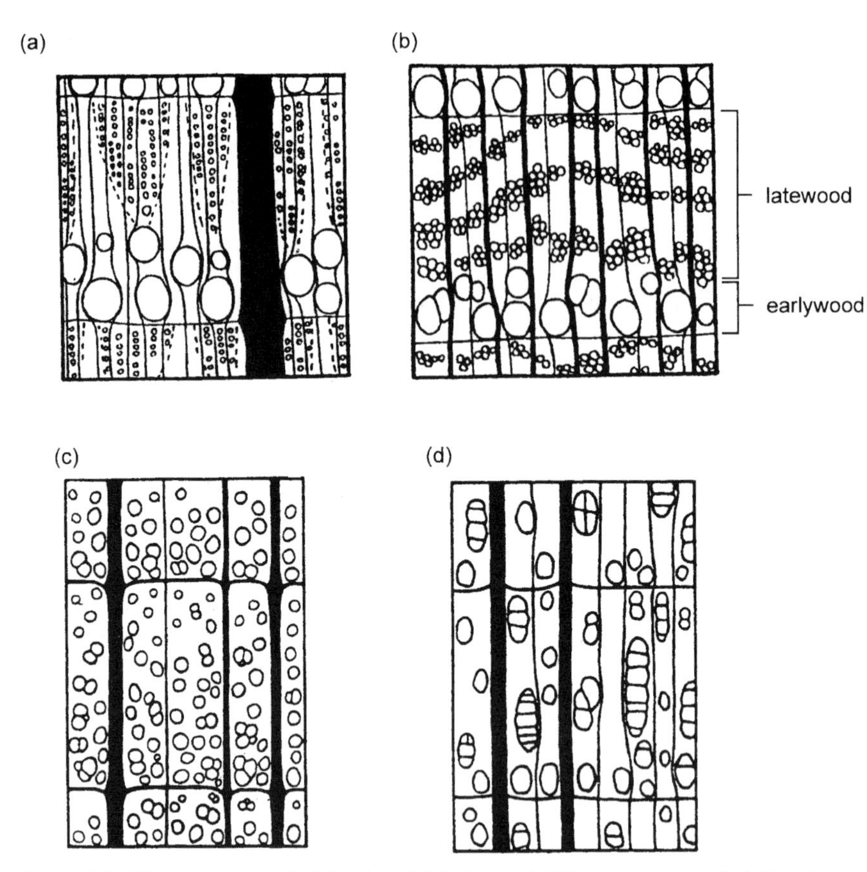

Figure 3.5. Ring-porous woods (a) oak and (b) elm; and diffuse-porous woods (c) beech and (d) alder. The centre of the tree is towards the bottom of the page; horizontal lines mark the edge of a year's growth; vertical lines are rays.

So, a growth ring starts off with less dense wood, which looks lighter from a distance, merging into denser and darker latewood followed by an abrupt join (representing winter) before the early wood of next year.

As is usual, things are a little more complicated in hardwoods! They can be divided into ring- and diffuse-porous trees. In ring-porous trees (like ash, elm, oak, black locust, hickory, catalpa and teak), the early wood is dominated by huge vessels, which contrast distinctly with the smaller ones of the late wood (Figure 3.5). Looking at the cleaned cut end of one of these trees, the large holes or pores in the earlywood can be seen to follow each ring around the tree: hence ring-porous. In diffuse-porous trees (like birch, maple, beech, poplar, lime, walnut, mahogany and eucalypts) the pores are more even in size and evenly distributed throughout the ring, making the rings less obvious. But the boundary between rings can still usually be seen as a thin dark line made up of small thick-walled cells that border on the earlywood.

Which trees are ring- and which diffuse-porous is not always clear-cut. It can vary within one tree; in a ring-porous tree, the innermost rings (i.e. the first rings grown), or ones grown while the tree is under great stress, can be diffuse-porous. Related trees, and even individuals in the same species, may be ring-porous in the north and diffuse-porous further south. Indeed, ring-porous trees are more common in mid-latitudes and diminish down to only 1% of species in the tropics. The ecological reason behind this apparent confusion is discussed in the next section.

In temperate areas, with a regular unfavourable period each year, we are well used to a tree forming one ring per year, enabling us to age a tree by counting the rings. Although this usually holds true, it is not always so. More than one ring may form in a year where growth has been interrupted by frost, fire, flooding, drought, defoliation, etc. and then resumed. Citrus trees have been seen to produce ten rings in one year! In other cases rings may be missing, sometimes from just part of the tree; yew is famous for its missing rings, sometimes hundreds. Localised damage to the cambium is one potential cause but the most usual reason is linked to the way a tree grows a ring.

In early spring the cambium becomes active at the base of swelling buds, spreading down the branches and trunk to the roots. In ring-porous trees this wave of new wood growth spreads rapidly downwards within days, giving the impression of being almost synchronous along the whole tree. In diffuse-porous trees the downward spread is slower, with new wood appearing at the base of trunk several weeks (sometimes almost two months in older trees) after its initiation in the twigs. Conifers are intermediate, usually taking about a week. The whole process is governed by hormones produced by the buds. Under difficult growing conditions (cold, drought) the ring may never make it to the bottom of the trunk, or may be restricted to one side of the trunk. Indeed, the incidence of missing rings at the base of a trunk increases with environmental stress. It is tempting to predict that missing rings should be more common in diffuse-porous trees and conifers. To make life more interesting, you should note that whereas earlywood starts from the top down, latewood begins at the base of the tree and moves upward, and ceases growing from the top down. Latewood is therefore thicker near the base of a tree and may disappear altogether in the upper reaches of a tree.

In the more equitable climate of the tropics, trees may grow continuously and so not produce rings, and even if they do, they need not be annual. In leaf-exchanging trees (Chapter 2), which may be leafless for just a few days, the cambium may not become fully dormant and so rings may be indistinct or missing. Half the species in the Amazon Basin produce no rings; in India the figure rises to three quarters.

Variation in ring width

Ring width varies tremendously owing to variations in the environment, rings getting narrower in years with less favourable weather. This is considered further in Chapter 6. Suffice it to say here that in conifers, as the ring gets narrower it is the earlywood that shrinks while the latewood varies less. So as the rings get narrower, the wood gets denser overall. Conversely, in ring-porous trees like oak the latewood gets proportionately less in narrow rings, so, because of the great vessels in the earlywood, wood with narrow rings is less dense. (The situation in diffuse-porous trees is less clear.) The conclusion is that to grow dense, strong oak, it should be grown fast while a conifer like pine should be grown more slowly. Indeed, most conifers in the UK (grown fast because of the warming influence of the Gulf Stream) are structurally unsuited for building, and such timber is imported into Britain from continental Europe (see Panshin *et al.* 1980 for further details).

How water gets up a tree

The tubes that carry the water up through wood are dead tubes. So how can water rise to the top of the tallest trees, over 100 m tall? There are two possible answers: it can be pushed up or pulled up. We will consider these in order.

Root pressure

In trees such as maples and birches a positive pressure builds up in the wood (xylem) several weeks before the leaves open in spring. Root cells (and sometimes cells in the trunk) secrete minerals and sugars into the xylem and water follows by osmosis, creating the positive pressure that forces water up the tree. Having said this, sap movement in maples is not by straight root pressure. The best flow occurs with warm sunny days above 4 °C and sharp frosts at night (below –4 °C). The exact mechanism is not clear but appears to involve compressed gas building up in the frozen xylem. Whatever the cause, can root pressure be used to explain the ascent of water in most trees? No, for several reasons. Firstly, many trees (e.g. pines) rarely if ever develop root pressure and in many deciduous temperate trees the root pressure disappears once the leaves emerge. Secondly, the pressure developed is rarely more than one to three atmospheres (0.1–0.3 MPa); enough to move water 10–20 m up a tree but clearly not enough to get water to the top of tall trees.

The question then arises, what use does root pressure serve? There are several potential answers. In trees and climbers of humid tropical regions where evaporation is very low, forcing water up through the plant ensures a supply of minerals to the top. Excess water is lost as beads of moisture that exude from the leaf tip and edges (guttation), and many species have special pores (hydathodes) at the ends of the veins for this purpose. I have a swiss-cheese plant (*Monstera deliciosa*) which produces large drops around the edges of the leaves in the morning; very unpleasant to brush past. A second reason is to refill tubes that have become air-filled (more later). Thirdly, it is a way in some temperate trees of getting stored sugars from the roots, up the tree, to supplement those moved through the phloem. The sugars are added to the xylem sap as the osmoticum. Thus, birches, maples and even the butternut of Eastern N America (*Juglans cinerea*) are tapped in late winter and early spring to collect the sugary sap (2–3% w/v sugar in maples, weaker in the butternut) to make wine and syrup (60% sugar). A birch may yield 20–100 l and the Canadian maple syrup output of 18.7 million litres in 1992 required close on a billion litres of sap!

Pulling water up trees: the tension–cohesion theory

It has been known since the work of Joly in 1894 that water is normally transported under tension, implying that it is pulled not pushed. Capillary action is probably the most obvious mechanism but is sufficient to get water only a metre up a tree. The mechanism to get water higher is quite simple in its elegance. The 'pump' in the system is the leaves. As the leaves lose water through their stomata (transpiration; see Chapter 4) and the internal cells become drier they pull water from the next driest cell and so on until the pull reaches a vein and exerts suction (or tension) on the water of the xylem. Because water is extraordinarily cohesive (it is a bipolar molecule with a positive and negative end, and this enables water molecules to cling together like magnets) the tension is passed on down the tree and has the effect of pulling a continuous column of water up through the xylem tubes. In fact, water is so cohesive that intact columns can be lifted over 450 m, more than enough for the tallest tree (currently 112.2 m (368 ft); see Chapter 6). Water can thus be moved up tall trees in copious quantities (see Chapter 2) by a combination of tension and cohesion: the tension–cohesion theory.

The estimated tension required to move water to the top of the tallest trees is about −30 atmospheres (−3 MPa). This is no mean tension. If the width of a tree is carefully monitored, the trunk is seen to get minutely thinner during a hot dry day as the suction pulls in the sides of the tree, and fatter during the night as it recovers. The fact that trees do not completely collapse under this enormous strain is a reflection of the strength of the individual tubes.

However, precise measurements have not always found these very large tensions. It is suggested (Canny 1995) that the living cells in the trunk (rays, axial parenchyma, and phloem) may exert a pressure on the xylem system: the compensating pressure theory. Thus the high tensions developing in the xylem would be lowered by the living cells pressing in on them in much the same way that a jet-pilot's suit inflates to squeeze his/her lower body to prevent all the blood pooling there during tight turns. In conifers this may be helped by the positive pressure that develops in the resin system. The controversy continues.

Air in the system

Sucking water up a tree only works if there is no air in the system. This can be demonstrated by using a thin plastic tube. If the tube is dangled in a bucket of water and suction is applied at the top, water will only rise about thirty feet (this takes some sucking!). No matter how much extra suction is applied the water will not rise further. This is because the column of water becomes so heavy that the extra suction simply acts to pull apart the air at the top of the tube, creating a partial vacuum. If the tube is full of water at the beginning, however, then water can be sucked up for hundreds of feet. This is why water pumps have to be primed and why tracheids and vessels begin life full of water: ready primed!

Trees, then, have to ensure that the columns of water remain free of air. A broken column can no longer lift water. Air can come from one of several sources: too much tension, or damage, or freezing. Under hot dry conditions leaves can lose water faster than the tree can take it up and there is increasing tension in the xylem (stomata in the leaves will close as a first line of defence; see Chapter 2). The tension can become so great that individual columns of water break with an audible snap; this 'cavitation' can take less than one hundredth of a second and produces the snap. Wider columns are more prone to cavitation, so, as you might expect, tube diameter is smaller higher up a tree where the tensions are highest. Root pressure can help refill tubes when it is available. Perhaps more useful is 'Mütch water'. When sugars are removed from the phloem in the roots and lower trunk, the water is recycled back into the xylem. Although this is only 1–3% of all water moved up the xylem, it is released primarily into the outer rings, where it may make up 45% of the total water content: enough to help refill cavitated tubes.

Winter is a distinct problem. Despite a widespread belief that trees drain in winter ('the sap going down in the autumn and up in the spring'), with the exception of some vines, they do actually stay full of water. As the water in the xylem cools and freezes it can hold less of the gas that is dissolved in it (ice holds 1000 times less gas than liquid water), which appears as air bubbles. In the spring these bubbles must be redissolved if the tube is to function. There are two distinct approaches to overcoming problems of air bubbles, the safe and the efficient.

In conifers and diffuse-porous trees the water carrying tubes tend to be narrow (see Wood structure, above). This means that they are less likely to cavitate during hot summer spells. Also, any bubbles forming in the winter are likely to remain small and in the spring will redissolve within minutes or hours with the aid of root pressure (although those trees without significant root pressure, such as poplars and beech, can take much longer). Thus, by the time the leaves start sucking hard, the xylem tubes are again water-filled. Here the apparently useless perforation plates and pits play a role; they catch bubbles when the ice thaws and prevent small bubbles coalescing into larger, longer-lived bubbles (the percentage of diffuse-porous species with complex perforation plates increases with altitude and latitude where it is colder in winter). In this way each individual tube may remain functional for several years and water can be taken up through a number of annual rings. This is the safety in numbers strategy: lots of narrow tubes each taking a comparatively small amount of water at a slow speed. This is ideally suited to cold climates where freezing is common. It obviously works because Italian alder (*Alnus cordata*) has been measured as losing more than 80% of hydraulic conductivity during the winter but by early spring the loss was down to less than 20% (i.e. the tubes had refilled) and remained under 30% during the summer.

The alternative to the conservative approach of the conifers and the diffuse-porous trees is seen in the more efficient but riskier throwaway approach of the ring-porous trees. They have wide and long vessels in the earlywood which allow the movement of prodigious quantities of water (making a tube four times wider increases the flow volume by 256 times, up to an upper limit of probably around 0.5 mm diameter[1]). Speeds of water flow up ring-porous trees is often of the order of 20 m per hour compared with 1–3 m per hour in conifers and diffuse-porous trees. But these superhighways of water are at great risk from air bubbles. Many of the vessels will cavitate under hot dry conditions, and the ones that do survive until the autumn are likely to develop large air bubbles during the winter which can take days or weeks to disappear in spring, effectively making them useless. In one study with red oak (*Quercus rubra*) 20% of vessels were embolised by August and 90% after the first hard frost. The solution to this problem can be found in how the trees start up in spring. In spring, ring-porous trees tend to leaf out later than diffuse-porous trees but if you measure trunk growth in early spring you find that ring-porous trees are already fatter with new spring wood. Ring-porous trees depend on the growth of the new earlywood in early spring (which starts ready filled with water) before the leaves emerge to provide the necessary water during the growing season. Thus most of the water moved up a ring-porous tree like oak is through the newest growth ring, which may be just a few millimetres wide.

Impressive through this is, it does seem very risky. Whereas a diffuse-porous tree may be sucking water through a dozen or so rings, a ring-porous tree is putting all its proverbial eggs in one basket. Or so it appears. Fortunately they have a good insurance policy in the form of small vessels in the latewood. If the big vessels of the outer ring become air-filled and stop conducting, the suction from the leaves becomes so great that water will move up the less efficient smaller vessels scattered in the late wood over several rings. It's a bit like an accident on a

[1] Experiments were once tried where coal gas was forced through branches and ignited at the other end. The length of branch through which this could be done gave an idea of the relative permeability to water: minimum lengths were, ash (ring-porous) branches 10 ft long; maple (diffuse-porous) 2 ft; in softwoods even 1½ in stopped gas flow. It is said that if you take a block of red oak 1 in × 1 in × 2 in and dip one end in detergent, you can blow bubbles with it!

motorway that forces cars on to smaller roads, which may not be as efficient but will get them to their destination eventually. A single ring, however, has important consequences for elms and Dutch elm disease (see Chapter 9).

Hydraulic architecture

The trunk is yet more complex. The leaves at the top of a tree are important because they are best placed in the competition for light. Because they are higher up, the water pathway is longer, requiring greater suction to get water (hindered further by the tubes becoming narrower to reduce the risk of air embolisms), and they may not always be the only ones in bright sun and so evaporating most water. The analogy is two children with straws in a milkshake; one straw is 20 cm long and the other 2 m; if they both start sucking the one with the shorter straw will get more than his share. So how can leaves at the top of a tree get their share of water when competing with those nearer the roots?

The answer is that a tree has constrictions at the junctions of the trunk and branches. These are created by a decrease in tube diameter, and, in hardwoods, by more frequent vessel endings. Lev-Yadun and Aloni, Israeli specialists in tree structure, also believe that the circular patterns that develop at branch junctions in softwoods and hardwoods, which contain non-functional circular vessels, increase the hydraulic segmentation of a branch from the stem (Figure 3.6). Conductivity through these junctions is often less than half that of the branch itself. This is like putting a crimp in the short straw to even up the share of milkshake. The mechanism is not perfect, however; when water is short, it is usually the top leaves that suffer first.

During drought, these constrictions may also prevent air embolisms that develop in branches from getting back easily into the main stem. In all trees, leaves and even branches are more readily renewable and it makes sense to sacrifice them in times of water shortage and keep the trunk working. Indeed, palms, which have only one growing point and one set of xylem tubes through their life, have considerable hydraulic resistance at their leaf bases.

Hydraulic network

Leonardo da Vinci considered a tree to be like a series of drainpipes where each set from the root gives water to a set part of the canopy. Although this idea has its uses (and has been championed by the Japanese in the Pipe Model of water uptake), it is evident that water-conducting tubes intermingle like spaghetti hanging from a fork and are abundantly linked together. This allows water moving up a tree to fan out around a growth ring (and to some extent between rings) with an angle of tangential spread of usually around 1 degree. Water from any one root spreads out as it goes up the tree, reaching not just one branch but a large part of the crown. This is a good safety measure since the loss of one

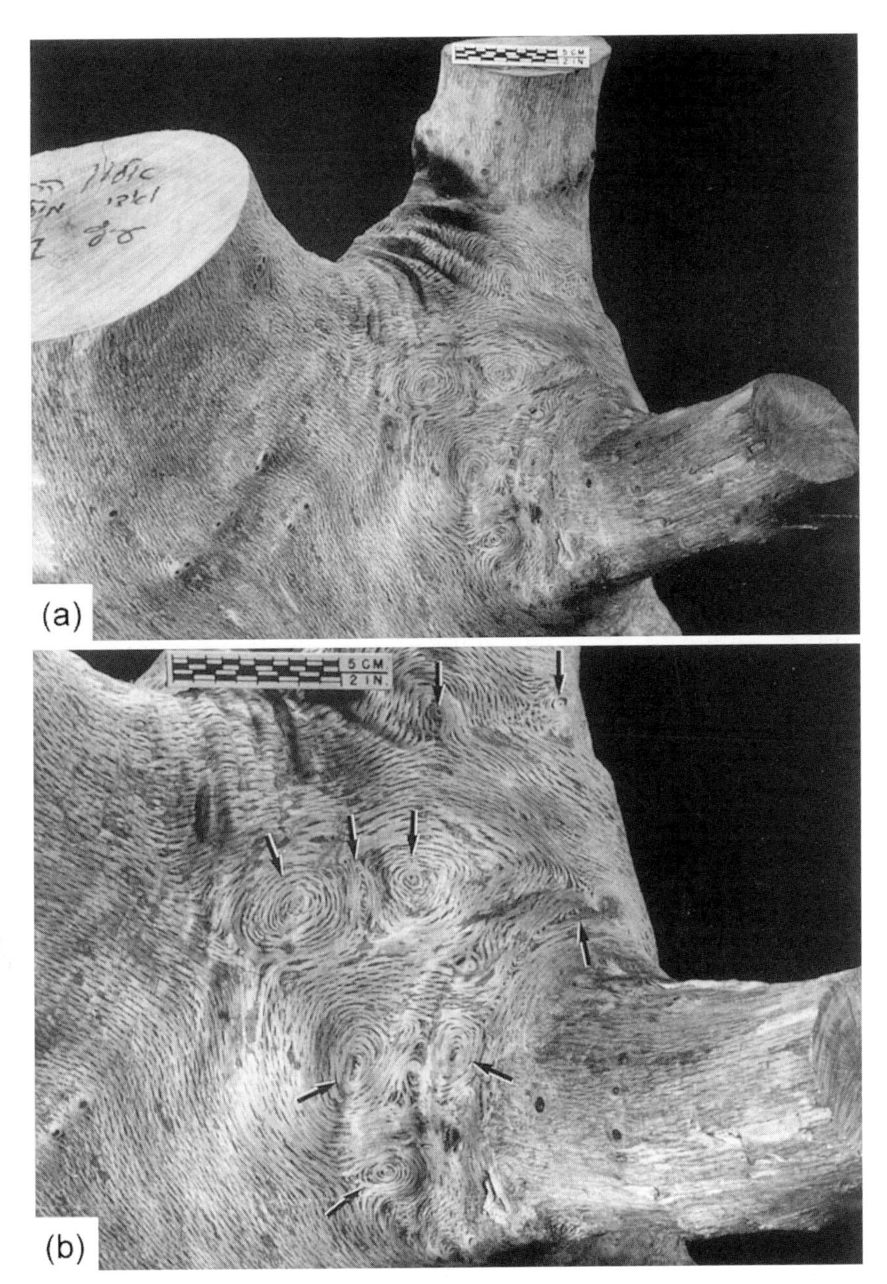

Figure 3.6. Branches from an oak (*Quercus ithaburensis*) with the bark removed, showing the non-functional circular vessels, which increase the hydraulic segmentation of a branch from the stem (arrows indicate the circular regions in the close-up photograph). These prevent a branch sucking up more than its share of water. From: Lev-Yadun, S. and Aloni, R. (1990). Vascular differentiation in branch junctions of trees: circular patterns and functional significance. *Trees*, 4, 49–54, Figure 1A, Page 50.

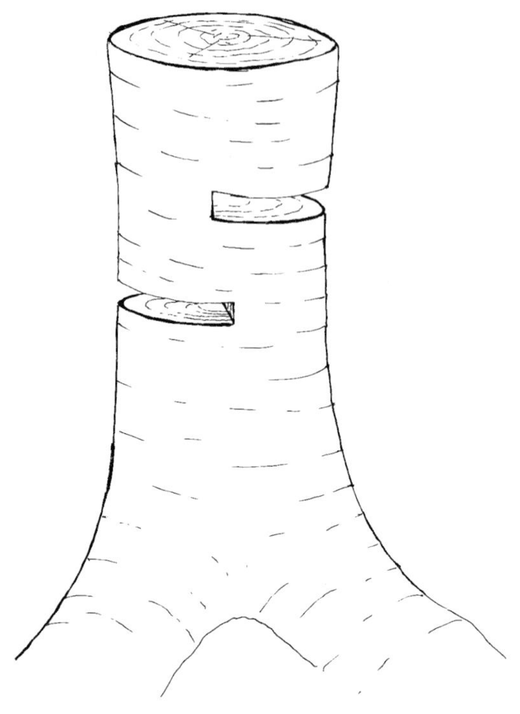

Figure 3.7. The famous double saw-cut experiment where two cuts are made one above the other from opposite sides of the tree. Despite the fact that all xylem tubes should have been cut, water can still be sucked up the tree provided the cuts are far enough apart.

particular root does not affect an individual branch but rather just reduces the overall water supply to the crown.

Such is the complexity of possible pathways of water up the trunk that water can still be sucked up under the most trying circumstances. This can be illustrated by the famous double saw cut experiments. Here two saw cuts are made into the trunk parallel to the ground, one above the other on opposite sides of the tree and overlapping, so that looking from the top, the tree is completely cut through (Figure 3.7). Yet water still ascends providing the cuts are more than a critical distance apart. One reason is that since the vessels grow crooked some may weave their uninterrupted way around the two cuts. More importantly, however, is how air is sucked into the cut vessels. When a vessel is cut by the saw the suction from above pulls air into the tube. But this air only goes as far up and down from the cut as the first complex perforation plate or end of vessel (which are rapidly blocked by tyloses or gums; see Chapter 9), or tracheid with its valve-like pits. Beyond these barriers the tubes remain full of water. So, providing there are water-filled tubes between the two saw cuts, water will be sucked up by being passed sideways between tubes. In short-vesseled trees such as maples, the cuts can be quite close (around 20 cm) but in oak, where the vessels are long with simple perforation plates, the cuts need to be more than 0.5 m apart.

Grain

The vertical orientation of most of the cells in wood is responsible for the 'grain' of the wood: cut across a tree and you are cutting across the grain. But the grain is rarely truly vertical, rather it spirals up the trees generally by just a few degrees off the vertical although it can be up to 40° or almost horizontal near the base of a tree. In a young softwood the grain starts off by usually spiralling to the left (like the middle stroke of the letter S placed on the bark), straightens up to almost vertical at around the age of ten before spiralling to the right (like the middle stroke of Z) at around 25 years of age. Spiral grain helps prevent a tree splitting between the rings (like pulling sheets of paper apart) under wind or snow load, and also appears to aid a more even distribution of water and food around the trunk and hence over more of the canopy and roots.

The desirable 'figure' or surface pattern of wood can be further influenced by the grain. In bird's-eye maple the grain grows in small pimples pointing in towards the pith which when cut through by the saw produce the pattern. Similarly, 'fiddle-back' in maples (commonly used for violin backs) is formed where the grain grows in vertical waves. Other features of the grain, induced by excessive strain, include internal cracks along the rays (usually called heart shakes) or around the rings (cup shakes, from the cupped pattern made when a plank from an afflicted tree falls apart).

Wood movement

Cellulose, the main constituent of wood, is a hygroscopic material that swells and shrinks as it absorbs and loses moisture. Wood in a living tree is usually always wet and so we only tend to notice this movement when the tree dies. Telegraph poles with excessive spiral grain can twist as they wet and dry, with imaginable effect on wires at the ends of the cross-arms. Cells shrink and expand in girth rather than length, so wood is fairly stable along the grain. In the same way, rays act as restraining rods reducing radial shrinkage, so most shrinkage is tangential (around the tree). A dried slice of tree nearly always has a 'pie-slice' crack running to the centre. Such change in dimensions is significant even over the normal temperatures and humidities expected over a year. For example, the pattern used for shaping the mould for the 16–17 t anchor heads of the cruise ship *QEII* is made of pine and the shrinkage between winter and summer (0.5–1.0%) makes a difference of 2–3 hundredweight (100–150 kg)!

Drying of wood can be significant to the living tree. This is where frost cracks come in, especially common in hardwoods such as oaks, planes and ashes but also firs (*Abies* spp.). Even in the relatively mild climate of Britain you can sometimes

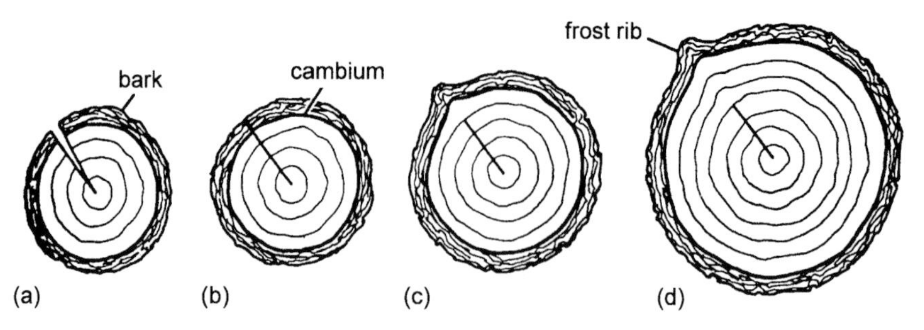

Figure 3.8. Frost cracks: (a) at the time the stem first cracked; (b) after the crack closed at the end of winter; (c) the crack covered by one summer's wood growth; and (d) covered by several years' regrowth, by which time enough wood has been grown to heal over the crack, leaving a frost rib down the trunk. Modified from Kubler, H. (1988). Frost cracks in stems of trees. *Aboricultural Journal*, 12, 163–75.

hear gunshot-like reports in the woods in winter as a tree suddenly cracks open. Once you get up off the ground, you can see what appears to be a drying crack running down the trunk (Figure 3.8a). And this is just what it is. On cold nights the outer part of a trunk freeze-dries in the same way that unwrapped food desiccates in a freezer. As it dries and contracts over the still-wet centre (aided a little by contraction of the wood due to the cold), the pressure is eventually relieved by the wood cracking open. Cracks on big trees in continental climates can be several centimetres wide in mid-winter. In the spring, the crack closes and the new wood formed that summer papers over the crack, but this may not be strong enough to resist cracking next year (the old wood in the crack never, of course, rejoins). Several mild years are usually needed to build a strong enough bridge over the crack to prevent future cracking. Repeated opening and healing results in the formation of protruding lips of callus along the edges called frost ribs.

Rays

We have seen that rays are strips of living tissue that run from the centre of the tree out through the cambium and into the phloem and so join the phloem and xylem into a giant circuit. One of their prime roles in the tree is to store food (they are also involved in the creation of heartwood; see below). When more food is made by photosynthesis than can be used straight away, and when minerals such as nitrogen are reabsorbed from leaves in the autumn, the surplus is stored in the living ray cells of the wood and bark (except the sieve and companion cells, where storage would interfere with function). The most favoured storage places are the wood of the roots and base of the trunk but even small twigs are used. Indeed, the sapwood of the sugar pine (*Pinus lambertiana*) from California and the Mexican white pine (pino de azucar, *Pinus ayacahuite*) both yield a sugary sap if damaged. Sugars transported by the phloem are converted into starch or fats when they arrive at the store. Around 1900 trees were classified as either starch trees (most ring-porous hardwoods plus some conifers like firs and spruces) or fat trees (most diffuse-porous trees plus pines) depending on what they primarily stored, but it is not always such a clear-cut distinction since both starch and fat can be found within the same species or even the same tree. Generally, low temperatures favour storage as fats.

As you might expect, an annual cycle of storage is most pronounced in deciduous trees. They have to grow a new set of leaves in spring before they can start making new food, and, as seen above, ring-porous trees have to grow a new ring of wood before that. The energy and materials for such new growth have to come from reserves held over winter, which consequently show a dip in spring. In most evergreen conifers (but not all, e.g. Norway spruce, *Picea abies*) the older needles from previous years are able to grow sufficient food in early spring to fund the new spring growth: storage tends to be much more equitable through the year.

Food is not stored just as a starter motor for spring. It is also an insurance against bad times such as losing parts of the canopy to gales, insect defoliation or rot. Trees generally maintain a steady reserve of stored food that is only used in times of great need; depletion of these reserves is a cause of tree death (Chapter 9). The reserves are also a savings account to fund years of abundant fruit production, especially in trees that show masting: see Chapter 5. Palms store impressive amounts of starch in the centre of the trunk to be used in flowering. This has not gone unnoticed by humans. The sago palm of South-East Asia (*Metroxylon sagu*) is felled, the central tissue is scooped out and the starch washed out and used as sago. Or the flower stalks are tapped for their abundant sugary syrup, which is used to make sugar and a potent alcoholic drink.

Sapwood and heartwood

A cut log often has a dark centre (the heartwood) surrounded by a circle of lighter wood (the sapwood; see Figure 3.1). The sapwood contains living tissue (as described above) and the rings that conduct water; the heartwood by contrast contains no living tissue and has materials added (polyphenols, gums, resins, etc.) which contribute to the darker colour. In some trees the heartwood is no different in colour from the sapwood, but is still there. Others do not seem to regularly form any heartwood at all; alder (*Alnus glutinosa*), aspen (*Populus tremuloides*), a variety of maples and other trees have been found up to 1 m in diameter with living cells and stored starch right to the centre (although some sugars found in, for example, pines, may be a by-product of heartwood formation). Fungal rot can add an extra confusion by producing discoloured wood that has nothing to do with heartwood formation! How heartwood is formed is discussed in Chapter 9.

What does heartwood do for the tree? Heartwood is usually denser and stronger than sapwood, and, because of the chemicals added to it, more resistant to rot (it's the part generally used for timber). It can therefore be seen as the strong core that holds the tree up. But this cannot be the entire answer because, as explained in Chapter 9, old hollow trees with little heartwood left may stand up to gales better than solid trees, and it is possible that trees deliberately court heart rot as a way of recycling all the stored minerals. Some might argue that heartwood is a very useful dumping ground for waste products that would otherwise be hard to get rid of. Against this must be set the knowledge that some compounds are expensively produced specifically to be incorporated into heartwood.

Sapwood is widest in vigorously growing trees but there are considerable differences between different species of tree. Northern temperate ring-porous trees, such as oaks, osage orange (*Maclura pomifera*), northern catalpa (*Catalpa speciosa*) and black locust or false acacia (*Robinia pseudoacacia*) normally keep just 1–4 rings of sapwood, which is generally lost with the bark when sawn for timber. Temperate diffuse-porous trees and conifers, on the other hand, can keep more than 100 sapwood rings, and tropical diffuse-porous trees such as ebony (*Diospyros ebenum*) may be practically all sapwood, with even a large tree yielding relatively little of the precious heartwood. This all makes sense if you remember that diffuse-porous trees spread their water uptake over a number of rings whereas ring-porous trees concentrate water movement in the outermost ring (see Growth rings above). But why keep more sapwood alive than necessary for water movement and food storage when the living tissue is a drain on the tree's resources? Heartwood, with no living tissue, has no such cost once made.

Heartwood does not necessarily follow a clear-cut cone inside the tree following the growth rings. In cross section the heartwood often meanders across rings, and generally the sapwood is thicker up towards the crown and gets thinner towards the base of the tree.

Bark

Bark is defined as all those tissues outside of the vascular cambium although, since the cambium comes away with the bark when it is stripped off, this may pragmatically also be considered part of the bark. (In a perverse way the cambium affects how 'glued on' the bark is: wood cut in summer when the cambium is actively growing tends to lose its bark upon drying, wood cut in winter does not.) Bark is made up of two portions, the inner bark or phloem (which conducts sugary sap around the tree), and the outer bark, which acts as the waterproof skin of the trunk, keeps out diseases and protects the vulnerable living tissue from extremes of temperature, even fire (look back to Figure 3.1).

Inner bark (phloem)

Movement of the sugars from the leaves where they are produced to where they are needed is quite different from the movement of water up through the wood.

The inner bark cells where this happens are thin-walled and alive. The active part of the phloem consists of living 'sieve cells' stacked one above the other to form tubes. The end walls are perforated (known as sieve plates) allowing sap to pass along the tube (Figure 3.4). In hardwoods each sieve cell is closely associated with one or more companion cells. The sieve cells do not have a nucleus, the source of genetic material necessary to keep the cell going (possibly because such a bulky object would impede the flow of sap), and it may well be that the companion cells provide support to keep the sieve cells functioning. In conifers the sieve tubes are more like the tracheids found in the wood in that they are discrete cells which overlap and communicate with pits rather than a sieve plate, and there are no companion cells. In both hardwoods and conifers the sieve cells are set in a matrix of strong fibres and parenchyma cells including rays. The strong 'bast' fibres of the inner bark have been (and are) used to make useful materials: Ugandans make cloth from *Ficus* trees, American First People made ropes and cords (especially from the linden tree, *Tilia americana*) and people of the Pacific islands, China and Korea made writing material from bark of the paper mulberry long before the pulping of wood.

Sieve tubes are not thickened with lignin like xylem tubes and so remain thin-walled. It is apparent that whereas the xylem tubes are built to withstand the great suction from above, the sieve tubes would just collapse under such negative pressure. Rather, the sieve tubes transport the sap under positive pressure. The exact mechanism is still being debated but the hypothesis with most support is one of 'mass flow'. Sugars produced by the leaves are actively pumped into the sieve tubes—this takes energy—which draw water into the tube by osmosis. This creates a high hydraulic pressure, forcing the sap along. At places where the sugars are used or stored, they are removed and again water follows by osmosis, lowering the pressure in the tube. With a high pressure at one end (the source) and low pressure at the other (the sink), the contents of the tube will move towards the low pressure. To give an idea of speed, the sap flow has been measured at 10 cm per hour in European larch (*Larix decidua*) to 125 cm per hour in American ash (*Fraxinus americana*).

The flow of sap does not have to be always down the tree, from leaves to roots. In spring, stored food can be transported upwards (along with sugar in the sap; see above), and developing fruits at the top of a canopy can draw food upwards from leaves lower in the canopy. One problem with a system under pressure is that any wound could cause the plant to bleed to death. Fortunately, the sudden release in pressure and rapid flow caused by a puncture causes a rapid sealing reaction at the pores of the sieve plates, isolating the damaged element. Flow then continues around the damaged element.

Sugars can be produced by photosynthesis during daylight more rapidly than they can be exported. The excess sugar is stockpiled as starch and then exported as sugar during the night. The transported sap contains primarily sugar but also small quantities of amino acids and other organic compounds (hence honeydew produced by aphids: the insects want these small quantities of nitrogen and have to filter through a lot of sticky sap to get them, and the waste syrup is expelled to drop on whatever is below, usually my car!).

If you look at a piece of bark taken from a tree, you will see that the inner bark (phloem) is comparatively thin (Figure 3.1). There are two reasons for this. Firstly, not much is made: typically 4–14 times more new wood (xylem) is made by the cambium compared with the phloem (although, xylem is much more affected by growing conditions than phloem and in bad years the width of phloem and xylem may be almost the same). The second reason is that the phloem is active for normally just one year (5–10 years, exceptionally in, for example, limes, *Tilia* spp.). Old phloem on the outside is crushed as the expanding tree meets the constraining outer bark, and is rapidly assimilated into new outer bark. In fact the layer of conducting phloem is usually less than a millimetre thick and has been measured at just 0.2 mm in American ash (*Fraxinus americanus*). Incidentally, growth rings appear in the bark in the same way as in wood but are a short-term record only.

Outer bark

The protective outer bark is composed mainly of old inner bark (phloem) that has been replaced by new working phloem from underneath. But cork is added to make the outer bark more weatherproof. To understand the process, let's start at the beginning. In a young stem, be it a seedling or a young twig on an older tree, the epidermis acts as the outer protective coat. Soon, however, the epidermis and green cortex are replaced by a corky layer (officially called the periderm; see Figures 3.2 and 3.9). A new growing zone (the cork cambium or phellogen) arises in the outermost layers of the cortex (occasionally in epidermis). This layer grows layers of cork (phellem) on the outside and occasionally a few cells on the inside, called the phelloderm. As the cork cells mature they are filled with air, tannins or waxy material and the walls are impregnated with a fatty substance (suberin) and sometimes waxes. The corky layer is therefore water- and gas-proof: superb as a weatherproof layer but not without its consequences. Firstly, it separates the epidermis from its food and water supply so it dies and scales away, which is a pity because the green cortex can add considerably to the photosynthesis of the whole tree (Chapter 2). However, in some species the phelloderm (the extra cells made on the inner side of the cork cambia) contain chlorophyll, allowing the corky bark to photosynthesise, especially where the bark is thin (such as in cherries and poplars). The second consequence of having a cork layer is that the exchange of gases across the bark is prevented. The living tissue in the bark, cambium and wood, like all living cells, need oxygen. This problem is solved by having holes in the bark, called lenticels, where the cork layer is ruptured into loose mounds of cells with numerous air spaces to allow gas exchange. These usually form beneath the stomata of the young twig. Lenticels are seen on

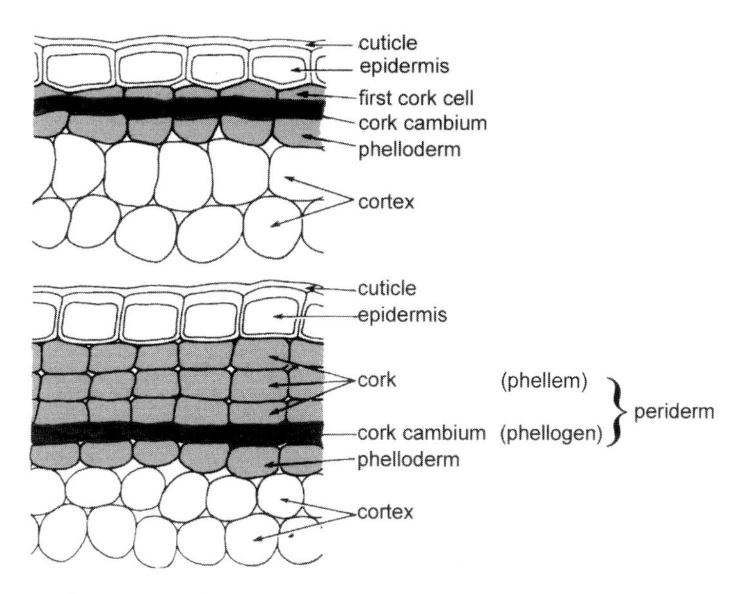

Figure 3.9. Cross-section of a piece of young bark showing the development of a cork cambium and the growth of corky cells. From Weier, T.E., Barbour, M.G., Stocking, C.R. and Rost, T.L. (1982). *Botany: An Introduction to Plant Biology* (6th edn). Wiley, New York. Reprinted by permission of John Wiley & Sons, Inc.

the bark as little pimples through to the horizontal raised lines on birches and cherries, covering 2–3% of the bark's surface.

As well as providing a weatherproof skin over the tree, the bark also protects against invasion from anything from bacteria to insects and mammals. In addition to the general toughness of bark, there are a number of chemicals found in some barks which have been shown to reduce attack. Some we find useful: quinine (from the bark of the S American quinine tree, *Cinchona corymbosa*, useful against malaria), the bark of the cinnamon tree of Sri Lanka (*Cinnamomum ceylanicum*), purgatives from buckthorn bark (*Rhamnus* spp.) and tannins for tanning leather from the bark of oak, birch, alder, willow, eucalypts, tropical mangroves (*Rhizophora mangle*) and a good many others.

Most monocots don't grow bark; instead, as in bamboos, the epidermis becomes very hard or, as in many palms, the existing tissue becomes corky beneath the surface. In those monocots with secondary thickening (such as *Yucca, Cordyline* and *Dracaena* spp.), a corky layer of sorts arises and may give rise to a very dicotyledon-like bark.

How does bark cope with the expanding tree?

In most trees, the phloem starts growing early in the spring anything up to 6–8 weeks before the xylem (although it appears more synchronous in some

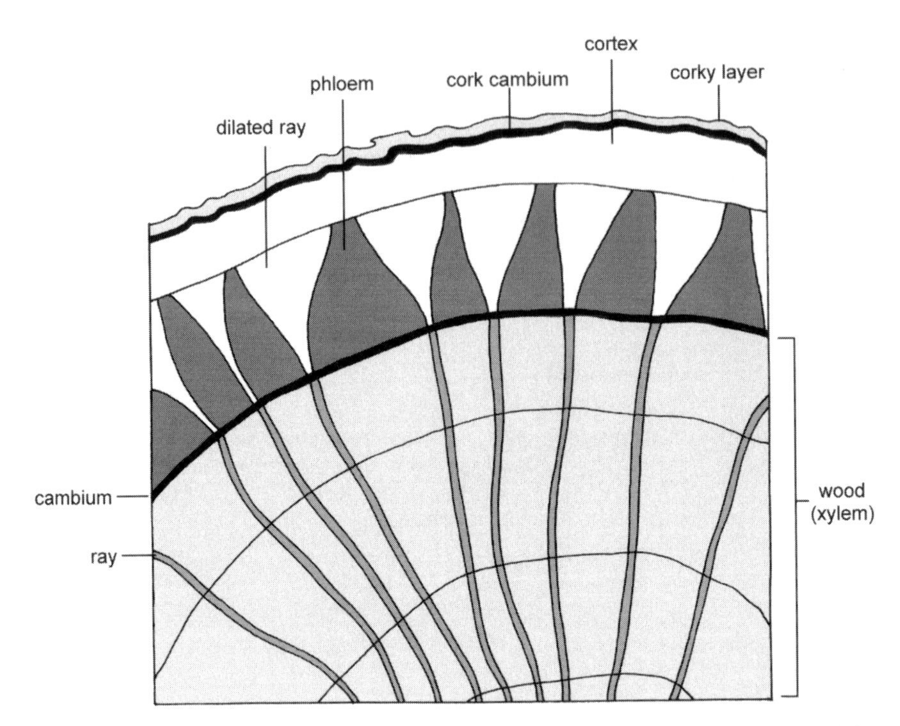

Figure 3.10. A diagrammatic view of a cross-section of a lime (*Tilia*) showing the expanded rays, which 'dilate' to help the inner bark (phloem) to stretch as new wood (xylem) grows under the bark.

ring-porous trees). So, as the new wood starts growing *inside* the bark, the newly formed phloem is immediately put under strain. An immediate solution is for the rays to grow sideways or 'dilate' so that they look like the mouthpieces of trumpets (Figure 3.10); this is like sewing a number of extra panels into a shirt that's too tight. Dilation growth is especially common in citrus and lime trees. But what happens as the tree gets fatter each year, and grows in diameter from a few centimetres to sometimes well over a metre, and the original phloem and bark get ridiculously too small?

A solution is to have lots of dilation growth, and indeed, in beech (*Fagus* spp.), by the time you get to the outerside of the bark it is also completely expansion tissue (Figure 3.11). This is helped by there being only a very small thickness of inner bark produced each year. The second element in this solution is the cork cambium. Some trees do not bother with producing cork from a cork cambium. Instead, the cortex and epidermis grow throughout the life of the tree to keep it covered, resulting in an effective but very thin bark, as in certain species of holly, maple, acacia, eucalypt and citrus fruits. In beech, however, there is a cork cambium, which normally remains active for the life of the tree and keeps up with the expanding girth of the tree by growing sideways. Very little corky tissue is

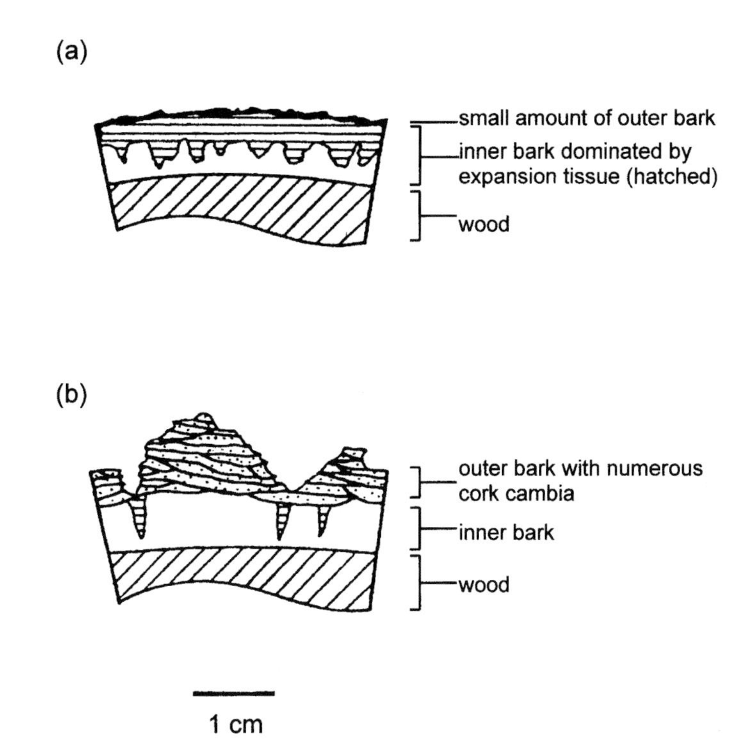

Figure 3.11. Bark cross-sections of (a) beech and (b) oak. Modified from: Whitmore, T.C. (1962). Why do trees have different sorts of bark? *New Scientist*, **312** (8 November), 330–1.

produced; the thin bark is shed in dust-like fragments, and the bark remains smooth. A similar type of growth accounts for the smooth bark of trees such as birches, cherries, hornbeams, some maples, and firs, although in these the original cork cambium may be replaced in later life (perhaps after a century) and accounts for the rugged bark found on old cherries, for example.

In most trees the inner bark grows four times or more faster than in beech and there is a great deal more tissue to be dealt with. Here, new cork cambia form successively deeper into the inner bark. These cambial layers usually have the shape of curved shells with their ends pointing outwards (producing plates) or longer lines down the bark, which with the strong bast fibres tends to produce the ridges seen in oak (Figure 3.11). These cambia may live for several years before being cut off and killed by new cambia below. The dead tissue of the outer bark cannot form expansion tissues so the outer bark splits as the tree increases in girth and changes from the smoothness of youth to the rugged pattern of maturity. The time taken for this change varies from ten years in Scots pine, twelve years in lime, twenty years in alder, to thirty years in oak.

An analogy for these two ways of forming bark is to consider what happens if you eat too much, expand rapidly in girth and your clothes get tight. You have two options if you want to stay covered up. The first is to have thin stretchy clothes similar to beech (but in beech the bark doesn't so much stretch as grow in diameter). The other option is to put on bigger clothes underneath the tight ones (not easy, but ensures continuous coverage) and let the tight ones burst off the outside demonstrating in their remains just how small you used to be. This is similar to how oak copes, where the old layers are left as ridges on the outside.

You can see from the above that bark consists mostly of dead phloem, often with many strengthening fibres, and a small amount of cork produced in thin layers. An exception to this is found in the cork oak (Figure 3.12). Like beech, the cork oak has a single cork cambium but here it is capable of producing a cork layer many centimetres thick over 8–10 years. When this thick, the almost pure cork is ready to be split from the tree by making a vertical slit in the bark and peeling off the cylinder of oak along the old cork cambium. The resulting pink trunk soon turns black and a new cork cambium forms deep in the old inner bark and produces new cork even faster than before, splitting the black skin. Each successive stripping of the oak produces better quality cork until by the third to sixth stripping it is good enough for wine corks. Remember that cork

Figure 3.12. Cork oak (*Quercus suber*), Birr Castle, Ireland.

layers cannot be completely impervious; they have lenticels to allow gases through the bark. Corks for bottles are cut, of course, so that the lenticels (those brown powdery lines) run *across* the cork to prevent spirits or other liquids escaping. Note that most trees die when the bark is stripped off around the tree because it separates from the tree along the *vascular* cambium (see Figure 3.2) and so all the inner bark is lost. In the cork oak the bark splits off along the *cork* cambium, leaving the inner bark in place.

When struck by lightning, oaks are seriously damaged more often than beeches. The reason is simple. When a tree is struck the charge follows the path of least resistance. With a thoroughly drenched tree this is down the water film on the outside of the trunk (little damage results) but in a dry tree it is through the water in the xylem, which causes the water to explosively boil. The rough bark of oak takes more rainfall to completely wet than does a beech, so oaks are more often at risk.

Bark thickness and bark shedding

Smooth barks can be surprisingly thin, being as little as 0.6 cm (6 mm) thick in a beech 30 cm diameter (compared with up to several centimetres in a similarly sized oak). Foresters go by the rule of thumb that bark makes up 7% of the volume of a felled beech log, and 18% in oak. This is not just a reflection of how quickly bark grows in oak but also how tenaciously the old bark clings on. In oaks, redwoods and other trees with thick barks, the successive cork cambia stick together so that the giant sequoia (*Sequoiadendron giganteum*), for example, can have a bark thickness of up to 80 cm or more, representing centuries of old phloem, creating a soft spongy fire-proof blanket around the tree.

On trees that *do* shed their bark, for example plane trees (*Platanus* spp.), the bark separates by tearing through thin-walled cork cells by the cork cambia, and so the shape and size of the shed pieces reflects the organisation of the successive cambia. Birches and cherries have alternating layers of thin- and thick-walled cork cells, and this results in the shedding of thin papery sheets. Some 'stringybark' eucalypts grow masses of thin-walled cells in the bark, so, when it dies, the bark breaks into strings of loosely attached fibrous bark. Shedding of bark is commonest towards the end of the growing season, and especially after hot weather, when shrinkage by desiccation helps pry loosening pieces off the tree. Why shed bark? In some cases it is a defence against things that would live on the bark. Certainly in California the Pacific madrone (*Arbutus menziesii*), a shrub with colourful orange bark which is regularly shed, is not parasitised by mistletoe as are surrounding trees. The London plane (*Platanus* × *hispanica*) has survived so well in London because it regularly sheds bark, taking with it lenticels blocked by pollution and exposing fresh open lenticels.

Trees grown in shade tend to have thinner bark, which explains why tree trunks suddenly exposed to the sun by the felling of neighbours can be damaged on the sunny side by sun-scorch, especially thinner-barked species such as beech, hornbeam, sycamore, spruce and silver fir. Transplanted trees are sometimes protected by being wrapped in paper, raffia or sacking.

Burs, buds and coppicing

Abnormal bulges or bumps can be found on the trunks and limbs of nearly any kind of tree. Although these burls or burs are common, we know very little about how they form. Insects, bacteria and viruses cause a variety of disruptions to the cambium and it may be that once started, the lumps enlarge as new wood is added each year, rather like a cancerous growth. Other burs are known to form where there are a mass of buds buried in the bark which distort the trunk's growth (Figure 3.13). These buds on the trunk (referred to as epicormic buds) are really dormant buds which formed as normal in the axils of leaves on the young shoot but which never developed any further. Instead they have remained dormant and have grown each year just enough to stay at or just below the surface of the bark (Figure 3.14). Since they often branch as they grow, a whole mass of buds may develop in one spot in the wood, causing the familiar bur. Indeed, if a bur is cut open the trace of the buds can be followed right back to the centre of the tree. You might think that all buds would open in the spring so these dormant buds would be quite rare, but it has been found, for example, that in red oak (*Quercus rubra*) about two thirds of buds remain dormant in

Figure 3.13. A giant bur on oak (*Quercus petraea*) that covers most of the trunk. Smaller burs appear as individual golfball- to football-sized swellings. Cannock Chase, Staffordshire.

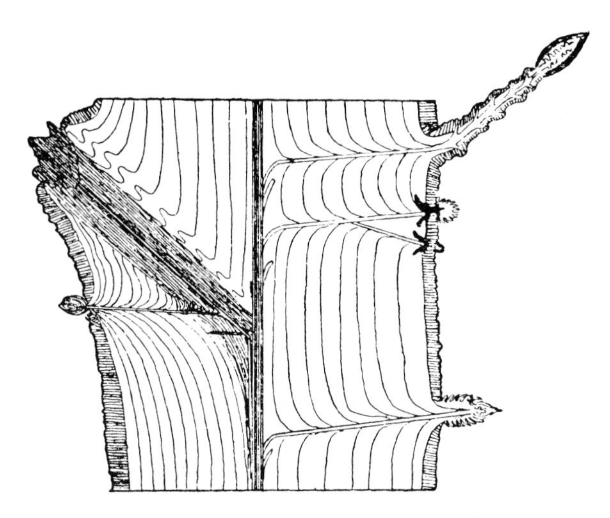

Figure 3.14. Longitudinal section of a tree trunk with epicormic buds arising at the centre of the tree and growing enough each year to stay at the bark's surface, often splitting as they go. From: Büsgen, M. and Münch, E. (1929). *The Structure and Life of Forest Trees.* Chapman & Hall, London.

spring. Many will be subsequently aborted (Chapter 7) but some survive. These dormant buds are the tree's insurance policy; if the crown of the tree is damaged (or we prune it) they will sprout to produce new branches. In trees like common lime (*Tilia* × *europaea*) and English elm (*Ulmus procera*) the buds are more exuberant and produce masses of shoots around the trunk's base even in normal healthy trees. Buds may be scattered widely up the trunk as in oak or concentrated at the base as in eucalypts and birch.

Epicormic buds can be 'adventitious', that is forming afresh from any living cells in the trunk, usually after some sort of injury. They are often formed on callus tissue covering a wound: new shoots on cut stumps of poplars, horse chestnuts, beech and walnuts are often from adventitious buds on the callus tissue rather than from dormant buds. Similarly, most root suckers are from adventitious buds formed when needed.

Note that a number of tropical trees and the common hazel (*Corylus avellana*) of temperate woods are self-coppicing and will readily grow new shoots from the base to produce the typical mass of shoots, although they tend to be more uneven in size than those resulting from a tree being felled.

Conifers are notoriously poor at growing from epicormic buds (stored or adventitious, but read about the formation of new buds in leaf axils in Chapter 6) and will die if cut to the base or are otherwise left with no foliage. There are exceptions, of course: the canary pine (*Pinus canariensis*), monkey puzzle (*Araucaria araucana*) and coastal redwood (*Sequoia sempervirens*) *will* regrow. Older hardwoods may lose the ability to regrow from epicormic buds. This has caused a problem in the UK with old pollards. Pollards are trees where the trunk is cut, usually at a height of 2 m (6 ft) or so, and left to regrow long straight wood of the sort much used when people lived off the land (coppicing is similar but is done at ground level; pollarding was common in parks where the new regrowth had to be kept out of reach of deer and other grazing animals). Conservation bodies would like to repollard historical trees that were last cut anything up to a century ago but such old branches have few buds left. Fortunately, a large body of research is accumulating on how to improve success (see Read 1996).

Branching and knots

Except for branches arising adventitiously (explained above), all branches can be traced back to the middle of the trunk or branch on which they are growing. As a new branch elongates in the first year, it lays down buds, which develop the next year into further young branches (this is examined further in Chapter 7). New layers of wood laid down each year coat the stick and its branches in a continuous layer (although rings are thicker on the trunk than the branch, so branch diameter increases more slowly than the trunk) giving the classic pattern seen in trees split along their length, seen in Figure 3.15a. The buried portion can no longer increase in diameter so it tapers to the middle of the tree (forming what is often called a spike knot). Shigo, an American forester who has spent his life dissecting trees, points out that reality is a little more complex. He maintains that new wood grows first along the branch and forms a downwardly turning branch collar (labelled (1) in Figure 3.16). As the trunk tissue begins to form later from above it produces a trunk collar over the top of the branch collar (2). The two collars develop water connections some way above and beneath the branch junction (4). Many times, however, the two collars will intermingle and not be visually obvious. The resulting collar at the base of a branch can often be seen as in Figure 3.17. As the trunk and branch get fatter, the bark tends to wrinkle in the crotch to form the 'branch bark ridge'. As the diagram shows, Shigo advocates pruning to leave the ridge and collar intact to ensure maximum callous growth and closure over the wound.

When we buy timber we are used to seeing knots in the wood. But why do some seem to be firmly anchored in the wood while others readily fall out? A knot is the portion of a branch buried in the wood as the trunk increases in diameter. If the branch was living, as in Figure 3.15a, the branch is continuous with the trunk and so the knot is held firmly and cannot fall out: an 'intergrown or tight knot'. If the branch dies, it ceases to increase in diameter and new growth simply surrounds the dead limb stump, so the branch is gradually buried as a cylinder of non-living tissue. When a plank is cut across this, it gives an 'encased or loose knot', often with bark entrapped: these 'black ring' knots are particularly prone to loosen and drop out (Figure 3.15b).

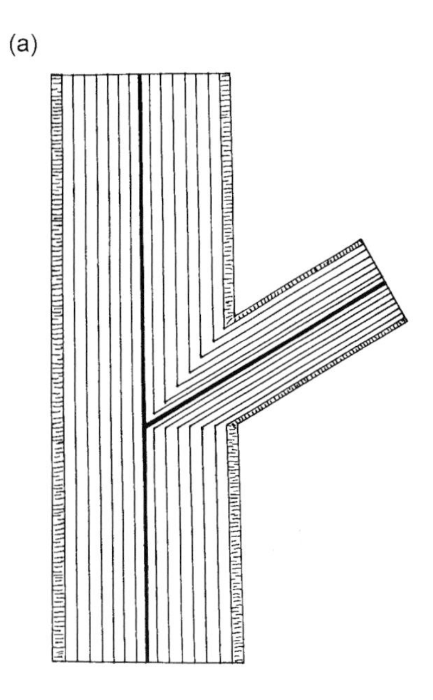

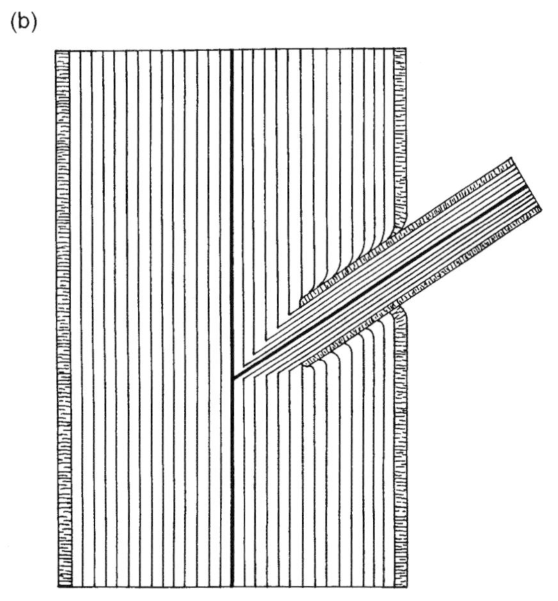

Figure 3.15. A split trunk shows (a) how growth rings cover the trunk and branch producing a 'spike knot' pointing to the tree's centre. (b) If the branch dies (after 6 years), then planks cut perpendicular to the branch will contain an 'intergrown knot' near the middle of the tree and an 'encased knot' near the bark. Intergrown knots do not fall out of a plank; encased knots can. Encased knots can be surrounded by bark or the bark may fall off before the branch is enclosed by new growth.

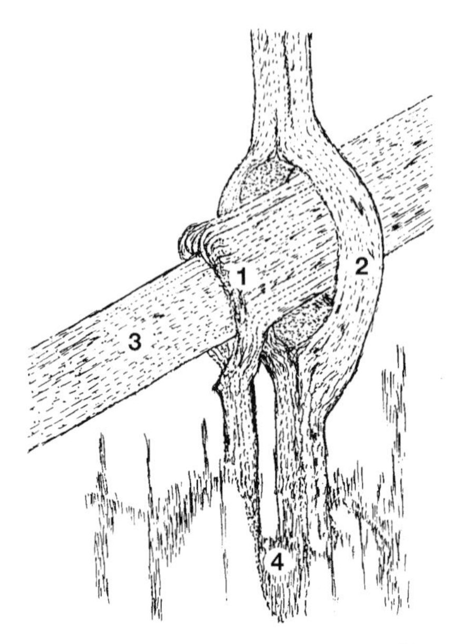

Figure 3.16. Branch attachment to the trunk. (1) The branch collar is grown in spring by the branch (3), followed later by wood growing down the trunk to form the trunk collar (2). The two collars develop water connections some way above and beneath the branch junction (4). From: Shigo, A.L. (1991). *Modern Arboriculture.* Shigo and Trees, Durham, NH.

Reaction wood

How do leaning trees straighten themselves? As is often the case, conifers and hardwoods go about it in opposite ways. With a few minor exceptions, conifers push and hardwoods pull. In a leaning conifer, 'compression wood' forms along the underside of the trunk in the new wood that grows after the tree starts leaning. This can be seen in a felled tree as a crescent-shaped darker zone of wider rings. As it forms, the compression wood expands and pushes the tree towards the upright. It is called compression wood because, obviously, it is in compression, and when cut out of a tree it does expand slightly in length. As you would expect, recovery to the vertical starts at the top (where the tree is thinnest) and gradually proceeds downwards (Figure 3.18). Often the top of the tree has returned to the upright while the lower part is still working at slowly bending the thick trunk, with the result that some years later the top of the tree ends up leaning the opposite way to when it started and it in turn grows more reaction wood to restraighten itself, leaving the tree with an S-shaped stem. Compression wood is readily seen in hemlocks (*Tsuga* spp.), which naturally develop drooping new terminal growth each year that by the end of the growing season has straightened itself (or nearly so) by using compression wood.

Compression wood is notably different from surrounding normal wood: the

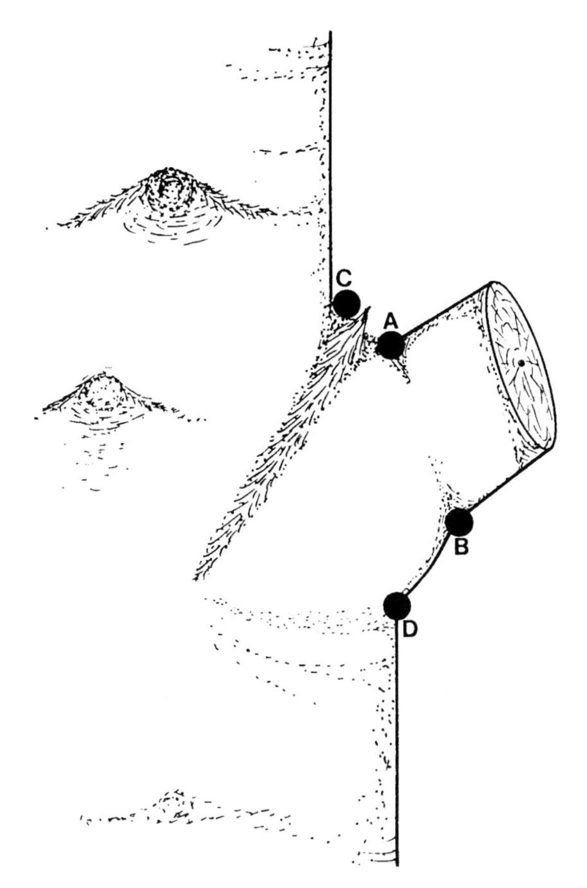

Figure 3.17. Correct pruning. Cuts should be made from A to B leaving the branch bark ridge (C) and the collar (B–D) untouched. From: Shigo, A.L. (1991). *Modern Arboriculture.* Shigo and Trees, Durham, NH.

tracheids are shorter and rounder with many spaces between, and they have a higher lignin and lower cellulose content, making the wood denser, harder, weaker and more brittle than normal with exceptionally high longitudinal shrinkage (ten times normal). Best avoided in woodworking, but bear in mind that compression wood can account for 20–50% of the wood in a Scots pine grown in a windy area.

Hardwoods, by contrast, form 'tension wood' along the *top* of a leaning stem, although it is not readily seen since it does not look superficially any different from normal wood. Nevertheless, it is there, contracting as it grows to pull the tree upright. The resulting wood is dominated by thick gelatinous fibres (principally crystalline cellulose with little lignin, the opposite of compression wood) with fewer and smaller vessels; this leaves the wood prone to splitting and cellular collapse on drying.

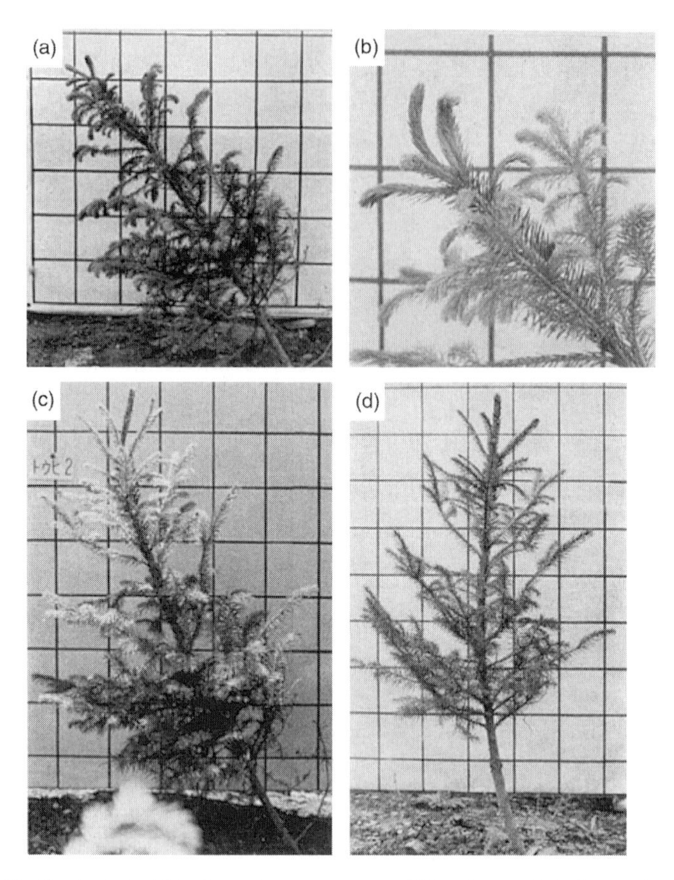

Figure 3.18. Gradual straightening of a young spruce (*Picea jezoensis*) in Japan. (a) Immediately after tilting the tree; the tree gradually straightens by the formation of compression wood on the underside of the trunk over (b) three days, (c) one month (with overcorrection of the tip) and (d) four months. From: Yoshizawa, N. (1987). Cambial responses to the stimulus of inclination and structural variations of compression wood tracheids in gymnosperms. *Bulletin of the Utsunomiya University Forests*, 23, 23–141.

Reaction wood is not just used to straighten leaning trees. It is used by trees to maintain their branches at the correct angle, and in some tropical trees there is evidence that tension wood is used to move the crown in an attempt to obtain sufficient light in dense forest. The controlling factor behind reaction wood appears to be the hormone auxin, although the exact mechanism is not clear. In a leaning stem, the normal flow of auxin down the tree is displaced by gravity and it accumulates on the lower side. And the same factor produces the opposite reactions in conifers and hardwoods.

Further reading

Atkinson, C.J. and Denne, M.P. (1988). Reactivation of vessel production in ash (*Fraxinus excelsior* L.) trees. *Annals of Botany*, **61**, 679–88.

Aloni, R., Alexander, J.D. and Tyree, M.T. (1997). Natural and experimentally altered hydraulic architecture of branch junctions in *Acer saccharum* Marsh. and *Quercus velutina* Lam. trees. *Trees*, **11**, 255–64.

Canny, M.J. (1995). A new theory for the ascent of sap – cohesion supported by tissue pressure. *Annals of Botany*, **75**, 343–57.

Cochard, H. (1992). Vulnerability of several conifers to air embolism. *Tree Physiology*, **11**, 73–83.

Hacke, U. and Sauter, J.J. (1996). Xylem dysfunction during winter and recovery of hydraulic conductivity in diffuse-porous and ring-porous trees. *Oecologia*, **105**, 435–9.

Holbrook, N.M., Burns, M.J. and Field, C.B. (1995). Negative xylem pressures in plants: a test of the balancing pressure technique. *Science*, **270**, 1193–4.

Kubler, H. (1988). Frost cracks in stems of trees. *Arboricultural Journal*, **12**, 163–75.

Lev-Yadun, S. and Aloni, R. (1990). Vascular differentiation in branch junctions of trees: circular patterns and functional significance. *Trees*, **4**, 49–54.

Milburn, J.A. (1996). Sap ascent in vascular plants: challengers to the cohesion theory ignore the significance of immature xylem and the recycling of Münch water. *Annals of Botany*, **78**, 399–407.

Milburn, J.A. and Zimmermann, M.H. (1986). Sapflow in the sugar maple in the leafless state. *Journal of Plant Physiology*, **124**, 331–44.

Panshin, A.J. and de Zeeuw, C. (1980). *Textbook of Wood Technology* (4th edn). McGraw-Hill, New York.

Pockman, W.T., Sperry, J.S. and O'Leary, J.W. (1995). Sustained and significant negative water pressure in xylem. *Nature*, **378**, 715–16.

Read, H.J. (1996). *Pollard and Veteran Tree Management II.* Richmond Publishing, London.

Rose, D.R. (1987). Lightning damage to trees in Britain. *Arboriculture Research Note* 68/87/PAT.

Sano, Y. and Fukazawa, Y. (1996). Timing of the occurrence of frost cracks in winter. *Trees*, **11**, 47–53.

Tognetti, R. and Borghetti, M. (1994). Formation and seasonal occurrence of xylem embolism in *Alnus cordata. Tree Physiology*, **14**, 241–50.

Whitmore, T.C. (1962). Why do trees have different sorts of bark? *New Scientist*, **312** (8 November), 330–1.

Zimmermann, M.H. (1983). *Xylem Structure and the Ascent of Sap.* Springer-Verlag, Berlin.

Chapter 4: Roots: the hidden tree

A common view of tree roots is that they plunge deep into the ground producing almost a mirror image of the canopy. Yet in reality a tree looks more like a wine glass with the roots forming a wide but shallow base (Figure 4.1). Most trees fail to root deeply because it is physically difficult and unnecessary. The two main functions of roots are to take up water and minerals, and to hold the tree up. In normal situations, water is most abundant near the soil surface (from rain), and this is also where the bulk of dead matter accumulates and decomposes releasing minerals (nitrogen, potassium, etc.). It should not be surprising, therefore, to find that the majority of tree roots are near the soil surface. The flat 'root plate' also serves very well for holding up the tree; deep roots are not needed (see Chapter 9).

Roots have other functions. They store food for later use (see Chapter 3) and they play an important role in determining the size of the tree. Roots normally account for 20–30% of a tree's mass (although it varies from as little as 15% in some rainforest trees up to 50% in arid climates). If the trunk is ignored (40–60% of the total mass), however, the canopy and the roots come out roughly around the same mass. This helps put into perspective the relative value of the roots and the leaves to each other. Too few roots and the canopy suffers from lack of water. Too few leaves and the roots get insufficient food. There has to be a balance. The roots 'control' the canopy partly through water supply but also by the production of hormones. If you doubt this control, think of fruit trees that are kept small by being grafted onto dwarfing root stock.

Water and dissolved minerals move only slowly through soil, so roots have to go to the water rather than waiting for the water to come to them. Although roots respond in a general way to gravity, they do not grow towards anything or in any particular direction; rather they are opportunistic in following cracks, worm runs, and old root channels, and in proliferating where they find water. Since soil is often very variable over short distances, root systems tend to be more variable in shape than shoot systems, and it is often hard to see any pattern in their growth. Nevertheless, the roots of most trees can be divided into a central root plate and an outer portion.

The root plate

Newly germinated seedlings produce a single root (see Chapter 8), which grows down as a young tap root. In some species (generally those with small seeds such

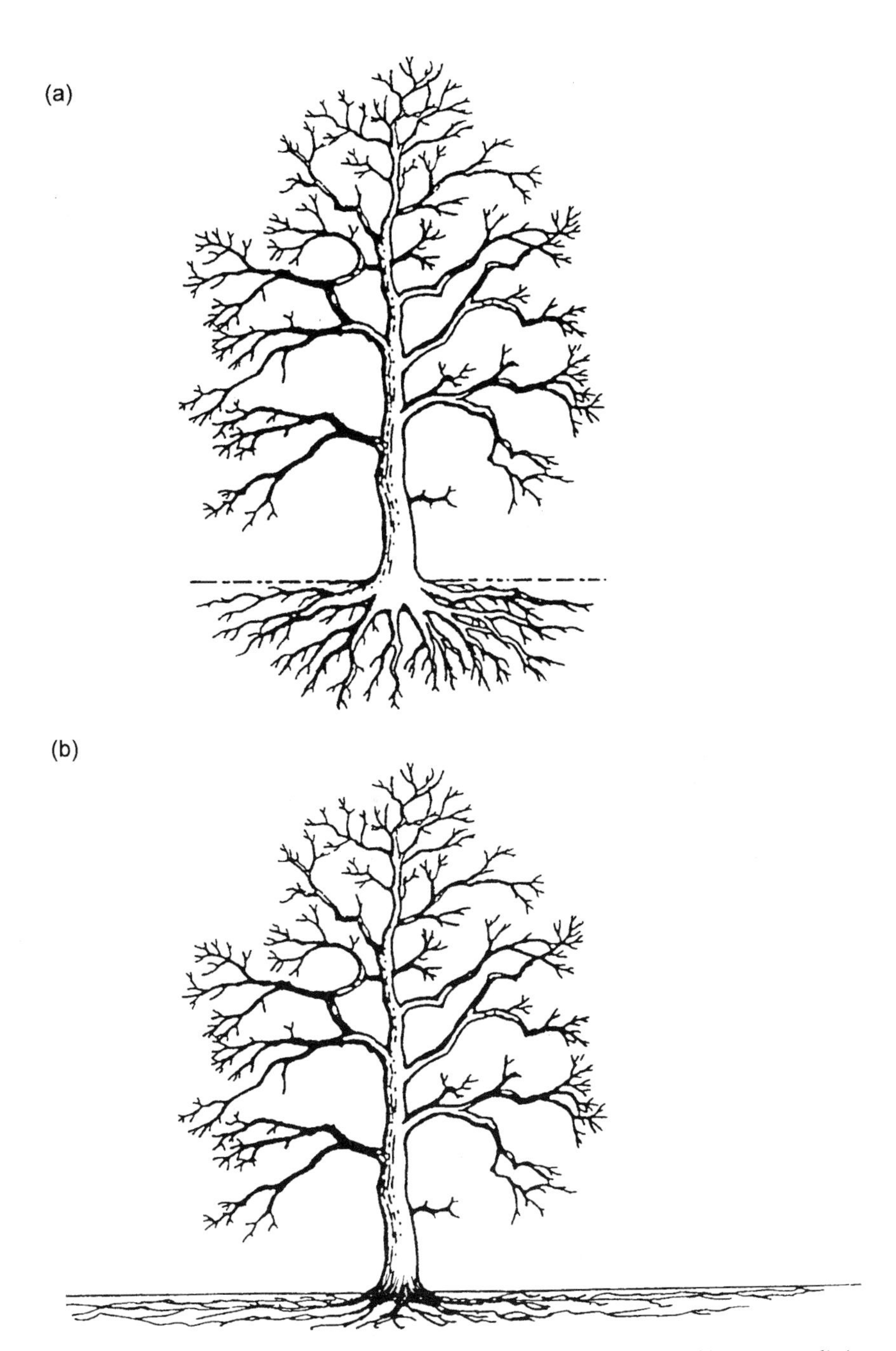

Figure 4.1. (a) The popular conception of what tree roots look like and (b) a more realistic representation. Based on: Helliwell, D.R. (1989). Tree roots and the stability of trees. *Arboricultural Journal*, 13, 243–8.

as spruces, limes, willows, poplars and birches) the tap root is small, easily deflected and seldom plays a major role; in a few species, such as willows, the tap root is so rudimentary that they are often described as having fibrous root systems. In other species, including pines, oaks, walnuts and hickories, the tap root grows more vigorously and initially dominates the root system; in oaks it may reach half a metre in the first season. But even here the tap root is not essential: trees from open ground sowings usually lose their tap roots when transplanted and do not suffer. In all but a handful of species the importance of the tap root is diminished as a number of major 'framework roots' (commonly 3–11) grow out sideways from the top of the tap root (Figure 4.2). These lateral roots may stay very close to the surface (as in spruce and fir) or, as in oaks, they may descend obliquely down several tens of centimetres before growing horizontally (or rather, parallel to the soil surface since roots can grow 'uphill'). Although the lateral roots can be well over 30 cm diameter at the base of the tree (helping to cause the marked swelling at the base of the trunk, the root collar), they quickly taper over the first metre to often less than 10 cm diameter, and down to just 2–5 cm over the first 1–4 m from the trunk, becoming less rigid and more rope-like. As these lateral roots grow out from the tree like the spokes of a wheel, they fork, branch and overlap. Within this criss-crossing framework the rigid large roots cannot move away from each other. As they grow fatter and so are pushed hard against each other, the internal tissues fuse together (aided sometimes by the swaying of the trunks wearing away the bark of crossing roots), creating a real connection between the roots (Figure 4.2b). This produces a solid root plate, the same width as the canopy or slightly wider, which moves as a unit as the tree sways and provides much of the anchoring and support for the tree. The root plate is made more solid and heavier still by the inclusion of rocks trapped between expanding, grafted roots. Darwin noted on the voyage of the *Beagle* that the inhabitants of the Radack Archipelago in the Pacific sharpened their tools by using stones prised from the roots of trees washed up on their beaches.

Deeper roots

In deep soils, trees may gain extra support and access to deep water supplies from sinker roots (also called striker or dropper roots), which normally grow down from the lateral roots within a metre or two of the trunk. Sinker roots can equal the length and thickness of the tap root, both of which grow down until they meet an obstruction, the water table or low oxygen concentrations. In temperate and tropical soils this leads to roots penetrating normally 1–2 m, and up to 3–5 m on well-drained soils (which debunks the myth that all tropical trees have shallow roots). Upon reaching their lowest depth the roots either grow horizontally to form a lower layer of horizontal roots (but generally not as vigorous or as branched as those near the surface) or they repeatedly branch to produce a bushy end like a broom (Figure 4.2a). To make the pattern more complicated

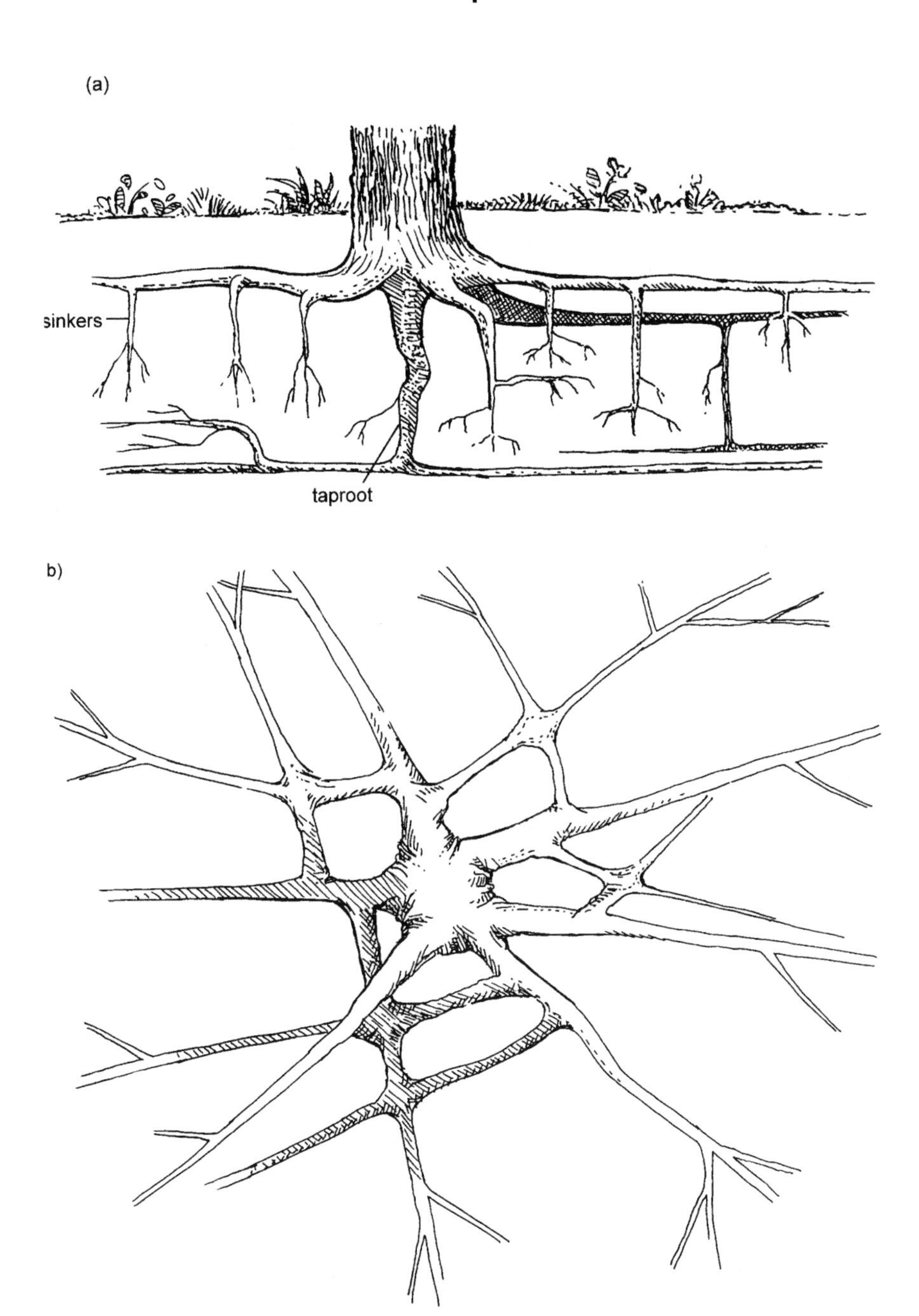

Figure 4.2. Diagrammatic view of the framework roots of a tree from (a) the side and (b) the underneath, showing a range of possible features.

still, in red oak (*Quercus rubra*) horizontal roots may grow obliquely upwards part or all of the way back to the surface before again growing horizontally!

Some trees are more capable than others of modifying their inherent growth pattern to suit soil conditions. For example, pines, which tend to have a long tap root and good laterals, can still grow on a range of shallow soils. Similarly, some willows and spruces, which have fibrous root systems and are normally at home on shallow or waterlogged soil, can root deeply on dry soils. In contrast, other trees including silver fir (*Abies alba*), sycamore (*Acer pseudoplatanus*) and oaks depend strongly on a deep tap root from which laterals are produced and do not do well on shallow soils. That is not to say that they are not capable of any modifications—droughted sycamore seedlings will grow deeper roots—but their plasticity is rather limited.

Although most trees do not root more deeply than a metre or two, there are many examples around the world of dry habitats where roots have been found to go much deeper to reach water: 12 m in an acacia unearthed in the excavation of the Suez Canal; 15 m in pines on the western slopes of the Sierras, USA. Deeper still were the roots of a mesquite bush (*Prosopis* species) found 53 m down in a gravel bed of an open-pit mine near Tucson, Arizona. At first the roots were thought to be fossilised and only after they were carbon-dated were they found to be less than six years old! In the deep sands of the Kalahari in Botswana, roots of a local tree (*Boscia albitrunca*) were found more than 68 m down a borehole (with the water table at 141 m). And fig roots in S. Africa have been reported at 120 m depth (400 ft). In these cases the roots grow deeply simply because in the easily penetrated, dry soils, they physically can, and this is where they can find water. In such extreme cases, where water is available primarily from deep water tables, the sinker and tap roots may replace the laterals as the main roots, producing the mythical rooting pattern expected in most trees. The dependence of 'normal' trees on the water table is discussed below.

Beyond the root plate

After a wind storm, leaning or fallen trees can be seen with the root plate lifted from the ground. This has undoubtedly led to the impression in many people's minds that tree roots spread only as far as the branches. Yet the lateral roots do not end at the edge of the root plate. Rather, these thin, and thus easily snapped, roots will explore the soil around the tree often for considerable distances. They appear unimportant: they are thin and rope-like (2–5 cm in diameter) with little taper but they can represent over half the mass and surface area of all the roots. Their role is not so much to hold the tree up but to search for water. At the edge of the central root system the lateral roots are already typically 1–2 m apart. To help fill the void between them as they spread out like the spokes of a wheel, they repeatedly fork. Forking seems to be primarily in response to injury or when the tip is forced to grow around an object: roots meeting a rock commonly fork either side of it. This may seem a very chancy process but is apparently reliable enough to work.

How far do roots spread?

In temperate trees the total spread away from the trunk is usually 2–3 times the radius of the canopy, and even up to four times the radius on dry sandy soils. Alternatively, root spread can be estimated as 1½–2 times the height of the tree. To put this into meaningful terms, the roots of an oak can reach up to 30 m away from the trunk (you should try pacing this out to appreciate how far it is). Underlying this is an inherent difference between trees. Some species such as European ash (*Fraxinus excelsior*) and the American southern magnolia (*Magnolia grandiflora*) have long, fast-growing, moderately branched 'pioneer' roots designed to exploit a large volume of soil; roots may be found up to four times the canopy radius away from the trunk. Others such as the European beech (*Fagus sylvatica*) and the American green ash (*Fraxinus pennsylvanica*) have short, slow-growing roots with many branches to utilise a small volume more effectively. This may explain why beech suffers in droughts; it uses up the available water and cannot exploit new areas of soil quickly enough.

. . . and how much damage can roots do?

Given that roots spread so far, and knowing that they can cleave rocks, should we be growing trees close to our homes? Fortunately, roots are only comparatively rarely implicated in damaging buildings. Direct damage to buildings, roads and paths can occur by expanding girth of the trunk and thick bases of the lateral roots. But once away from the trunk, roots do not thicken appreciably and tend to grow around obstacles given the option. These roots can sometimes still lift lighter structures such as paths and driveways but this is most likely by those trees with an abundance of surface roots such as ash, cherry, birch and pine.

Roots are not very good at penetrating compact or hard substrates and are therefore easily stopped unless they can follow a line of weakness such as an old root channel or a crack in your house foundations. Once in, though, they can exert considerable sideways force as the roots grow. This can be enough to cleave rocks along bedding planes and displace masonry in an impressive show of strength. The paradox of these strong yet weak roots is often seen in planted trees. Smooth compacted sides in a dug hole, especially in clay soils, cause the new lateral roots to grow around the inside of the hole, producing 'girdling roots' that wrap themselves around and around the trunk (especially prominent in maples for some reason). The roots may never escape from the planting hole—with implications for stability and water uptake—and if the tree survives long enough, the girdling roots can restrict the expansion of the trunk and seriously hinder growth. The same problem can be caused by roots growing around the inside of a container; once they start they tend to continue even when planted out. In commercial conifer growing, the small containers used have ridges to break the pattern, or they can be coated with copper sulphate: when the roots reach it they stop growing. Root pruning can cause similar problems when new lateral branches (at right angles to the original root) grow tangentially across the trunk.

Tree roots can be readily stopped by various permeable barriers made of plastic fabric or copper mesh. When buried in the soil, between, say, a tree and a driveway, these barriers allow through water but not roots. The take-home message is that, provided your house foundations are sound, tree roots should cause you few problems. Unless your house is on clay, that is.

A common problem is building subsidence due to shrinkage of clay soils caused by trees extracting water. In a survey in southeast England, Cutler and Richardson (1981) found that on clay soils oaks and poplars damaged buildings 30 m away from the trees, and willows 40 m away (Box 4.1). Oaks and poplars were particularly menacing because they caused damage out of all proportion to their abundance. Other large trees such as ash, elm, lime, maple, and horse chestnut were found to have damaged buildings more than 20 m distant. Even *Prunus* and *Sorbus* species and fruit trees (all Rosaceae) could cause subsidence over 10 m away. The good news, if you live on similar clay soils, is that the roots of the increasingly common cypresses, including Leyland cypress (× *Cupressocyparis leylandii*), spread a long way (20+ m) but 90% of the recorded damage was within 5 m of the tree. Felling a large tree on clay soils can be equally damaging as the water previously used by the tree causes the clay to expand.

Perhaps the most widespread problem caused by tree roots is the penetration and clogging of drains and sewers, which, being warm and wet, must be root paradise! And they can be big: one champion willow root taken from a storm sewer in Utah by RotoRooter Corporation was 41 m long (Figure 4.3). Roots will proliferate around pipes that have condensation on the outside or are leaking, and where they find an opening or crack they waste no time in taking advantage of it. Poplar, willow, horse chestnut and sycamore tend to be particularly invasive.

Fine roots

The large woody roots of the root plate and outer root systems dictate the overall size and shape of the whole root system, but it is the plethora of increasingly fine roots that provide the really intimate contact with the soil. These fine roots are often called 'feeding' roots but it is better to avoid this term. Tree 'food' is manufactured by the leaves as sugar; roots take up water and minerals, which are used by the tree in conjunction with these sugars to live and grow.

From the radiating lateral roots, small-diameter woody branches grow outwards and upwards, often reaching several metres long, branching four or more times to end in fans of short, fine non-woody roots. In red oak (*Quercus rubra*) there may be as many as 10–20 of these branches along a metre of lateral root

Box 4.1. Distances over which trees have been seen to cause damage to buildings by subsidence on predominantly clay soils in south-east England

These are the results of a survey from 1971 to 1979 conducted by the Royal Botanical Gardens, Kew (Cutler & Richardson 1981). Dashes indicate that data are not available.

Common name	Latin name	Maximum tree-to-damage distance (m)	Distance within which 90% of damage cases were found (m)	No. of trees
Willow	*Salix*	40	18	124
Oak	*Quercus*	30	18	293
Poplar	*Populus*	30	20	191
Elm	*Ulmus*	25	19	70
Horse chestnut	*Aesculus*	23	15	63
Ash	*Fraxinus*	21	13	145
Lime	*Tilia*	20	11	238
Maple	*Acer*	20	12	135

Box 4.1 *(cont.)*

Common name	Latin name	Maximum tree-to-damage distance (m)	Distance within which 90% of damage cases were found (m)	No. of trees
Cypresses	*Cupressus, Chamaecyparis*	20	5	31
Hornbeam	*Carpinus*	17	—	8
Plane	*Platanus*	15	10	327
Beech	*Fagus*	15	11	23
False acacia	*Robinia*	12	11	20
Hawthorn	*Crataegus*	12	9	65
Rowan + whitebeam	*Sorbus*	11	10	32
Cherries, etc.	*Prunus*	11	8	144
Birch	*Betula*	10	8	35
Elder	*Sambucus nigra*	8	—	13

Box 4.1 *(cont.)*

Common name	Latin name	Maximum tree-to-damage distance (m)	Distance within which 90% of damage cases were found (m)	No. of trees
Walnut	*Juglans regia*	8	—	3
Laburnum	*Laburnum*	7	—	7
Fig	*Ficus carica*	5	—	3
Lilac	*Syringa vulgaris*	4	—	9

in the root plate area, dropping to one branch every 1–5 m in the outer root system. Side roots tend to stay thin, seldom over 4–6 mm diameter in red oak: they are there to hold the finest roots, not to hold the tree up.

The finest roots are typically 1–2 mm long and 0.2–1.0 mm in diameter but can be down to 0.07 mm in diameter. Since they are easily broken with a trowel they usually go largely unnoticed. The painstaking work of Lyford (1980) sheds light on the impact of such fine roots. He dissected 1 cubic centimetre cubes of soil taken a few centimetres down into the floor of a red oak stand in the Harvard Forest, Massachusetts. In these small cubes he found an average of 1000 root tips, more than 2.5 m of root with a surface area of 6 cm^2 (six times the area of the top of the cube) not counting mycorrhizas (see below) or root hairs. Although this may sound as if the soil must have been solid root, most of the roots were so fine that they actually filled only 3% of the cube volume. Scale this up and Lyford has suggested that a mature red oak may have 500 million live root tips. Reynolds (1975) estimated that in Douglas fir, half the total length of all the roots were less than 0.5 mm thick and 95% less than 1 mm.

We have already seen that tree roots as a whole tend to be near the surface of the soil. In typical clay–loam soils 99% of roots are in the top 1 m and most are less than 20–30 cm below the surface. But if we consider just fine roots, the shallowness is even more impressive. In a N Carolina oak forest 90% of the mass of

Utah Operator Snags 114½ Ft. Root

This willow root, 114½ feet in length, was taken from a storm sewer recently by Provo, Utah franchise owner Don Wismiller. The giant was the largest ever encountered by Mr. Wismiller in his nine years of experience in the Roto-Rooter business. Even this, however, was not the complete root, for an obstacle was hit and the root chopped off at that point, leaving an additional twenty feet or so to the tree.

The line was part of a drainage system designed to prevent basement flooding in a comparatively low residential area. Many of the residents, however, had reported as much as a foot of water standing in their basements during the last year. Various efforts had been made to stop the seepage, but to no avail. Some had thought the water was coming from excessive drainage caused after the devastating floods last spring, since the homes were close to a creek.

At a neighborhood meeting which included the Council and the Mayor, the decision to call Mr. Wismiller was made after an examination of the plans of the installation revealed that the line ran under a huge willow tree in one of the yards.

Mr. Wismiller's speedy work quickly satisfied all the residents of the area, of course, and all agreed that the giant root which nearly filled the eight-inch line was the real source of their trouble.

Shown above, holding the culprit, are two of the neighborhood children. A similar picture was included in a writeup of the incident which appeared in the Provo Daily Herald.

roots less than 2.5 mm diameter were in the top 13 cm of soil. In an acidic beech forest of Germany it has been calculated that 67% of roots less than 2 mm in diameter were in the top 5 cm and 74% in the top 7 cm. In tropical trees the pattern is less clear but fine roots are probably even more strictly limited to the upper 5 cm of soil and often form a mat of roots over the soil surface. In Costa Rica, for example, this root mat can contain a third to a half of all the fine roots.

In a woodland situation these very fine roots are actually growing high up in the leaf litter, often in the moist compacted litter below the last leaf fall. Here the very small roots penetrate between the layers of leaves primarily by force (but sometimes by dissolving a path), branching and spreading in the same plane as the laminated leaf litter. It follows therefore that in woodlands, fields and gardens the finest roots are among the grass and herb roots, competing side by side for water and minerals. Thus our view of tree roots as being large and deep needs to be replaced with a picture of a stout central framework with radiating rope-like roots complemented by an incredibly delicate system of small woody and non-woody roots very close to the soil surface.

Having said that fine roots are very shallow, it is worth noting that some will always be found deeper, associated with the structural roots. To use the Costa Rica example again, around 5% of the fine root mass and 13% of all roots can be found 85–185 cm deep into the soil. The fact that there are some deep roots has been utilised in agroforestry where trees and crops are grown together with the assumption that the crops are using water near the soil surface and the trees are using deeper water, maybe 20–30 cm down. However, this assumption has never been proved and inevitably there is some competition between the crops and trees; crop yield is normally lower near the trees.

Fine roots and tree health

The fact that fine roots are close to the soil surface has several implications for looking after trees. When fertilising trees there is no need to dig pits or use 'tree spikes'; research has found no difference in the response of trees to fertiliser placed in holes versus that scattered on the surface. The roots of a tree and grass are competing in the same space for the same things, which explains why grass and weed control around newly planted trees is so important (a grass-free circle 2 m in diameter can dramatically increase root development in a seedling). In the same way, when we spread herbicides on our lawns or on paths and driveways (where it seeps into joints and cracks) we can also be killing tree roots. This may not seem too important but imagine a tree growing beside a road as in Figure 4.4. Most of the fine roots will be in the front lawn because that's where most water and nutrients will be available. Herbicide on that lawn could be

Figure 4.3. An illustration of how long roots can grow in a storm sewer. This willow root (41 m long) was reported in the magazine of the RotoRooter Corporation, Iowa in 1954.

devastating to the tree. How many gardeners would worry about an oak tree 15 m away? And yet as we have seen, the roots can spread over 30 m. For a similar reason, digging in the garden is another common cause of tree-root death. You might ask at this point how lone trees can then survive surrounded by a ploughed field. The answer is that with repeated disturbance the roots proliferate below the level reached by the plough or spade. The problem comes when an area is dug or ploughed for the first time, damaging a high proportion of the roots.

Figure 4.4 raises several other potential problems for urban trees. The spread of roots can be seriously affected by the impervious surfaces of roads and courtyards. Open soil can be no better when it is compacted by feet (and pigeons are a prime offender since their small feet exert more pressure per square centimetre than heavy machinery). Compaction seals the surface, preventing gas exchange and water penetration, and makes root penetration impossible. Fortunately, many trees can brave the low oxygen and water content beneath these barriers, pushing roots through to more favourable areas (trees such as cherries are known for producing root suckers on opposite sides of roads) or following the cracks between paving stones. Others, exemplified by the Japanese katsura tree (*Cercidiphyllum japonicum*), cannot apparently tolerate a paved surface over their roots, which must restrict root spread with obvious consequences in a drought or high winds. Perhaps the biggest problem comes when the area around a previously free-growing tree is paved: many of the existing roots die as oxygen is consumed in the soil, leading to an uncertain period when the life of the tree is in the balance before new roots can grow into new areas rich in oxygen and moisture.

Similarly, if the soil level is raised around established trees as part of landscaping (or naturally by the addition of silt by flooding) ill health and death can follow. Sinker and deep horizontal roots are likely to be starved of oxygen but, perhaps more significantly, the fine roots end up below the zone where most water and minerals are available unless the added material is very porous. If you run a playground or like to mulch your garden you'll be pleased to know that even 50 cm of wood chips is sufficiently porous to cause little problem for tree roots. Some trees (notably willows) can cope with extra soil by producing new roots on the buried stem, within days of burial.

Underground utilities can be devastating for tree health. Cables and pipes are usually buried deeper than the roots: the trouble comes in putting them in. A trench dug along the inner side of the path in Figure 4.4 would neatly sever most of the main roots of the tree, making it much more likely to die of drought or be blown over. So why don't we see rows of dead trees along streets if trenching is that bad? The answer is that sick urban trees are usually removed branch by branch or are felled long before they get to the stage of being an obvious danger. We usually notice the problem when we realise that the trees are missing. But there are moves afoot to help our beleaguered urban trees by sympathetic handling, as you can see in Figure 4.5.

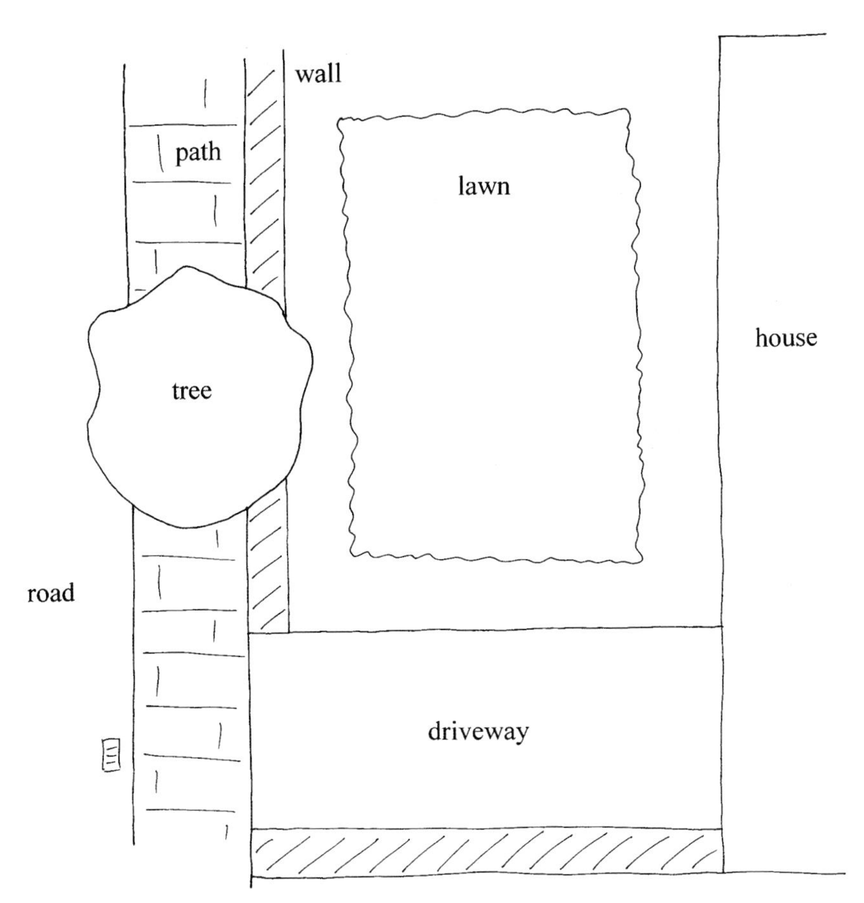

Figure 4.4. A hypothetical tree growing in an urban setting. Inevitably most of the tree's fine roots, and many of the larger structural roots, will be beneath the lawn. This makes any trenching operations along the inner side of the path much more serious for the tree than if it was in an open field.

Root loss and death

Root loss raises the question of how many roots can be lost before a tree suffers. This obviously depends on several factors but as a general rule many practitioners would consider the risk to be small if roots are cut off beyond the edge of one side of the canopy. A tangential straight-line cut along the edge of the canopy would cut off about 15% of roots. If the straight-line cut is made midway between the edge of the canopy and the trunk then around 30% of roots will be severed and trees of reasonable health, with roots previously unhindered in any direction, should be able to survive. In practice, 50% of roots can sometimes be removed with little problem provided there are vigorous roots elsewhere. Inevitably this degree of root loss will temporarily slow canopy growth and even

1

THE LIFE OF THE TREE IS IN YOUR HANDS

Even the biggest trees have nearly all their roots just below the surface: they're the trees' life support system. If the roots get chopped, the tree will be damaged and may **die!**

2

PLAN WELL AHEAD

Street trees need special treatment-cable trenching should be planned well ahead.

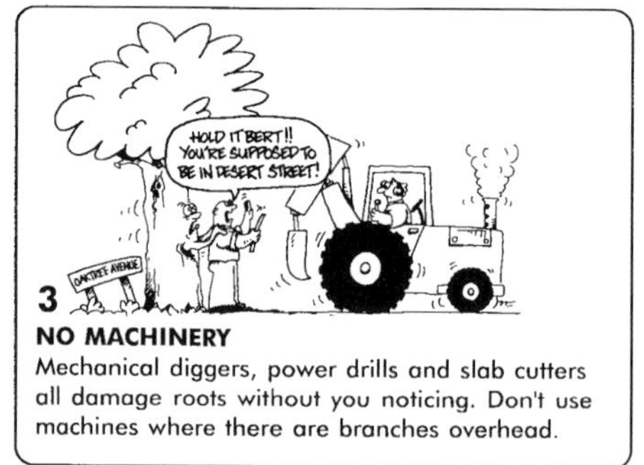

3

NO MACHINERY

Mechanical diggers, power drills and slab cutters all damage roots without you noticing. Don't use machines where there are branches overhead.

4

DIG CAREFULLY BY HAND

Take care with roots thicker than your thumb. Don't break or scuff the roots. Don't let the roots dry out: cover them with damp sacking or spray with water.

5

SLIDE THE DUCT UNDERNEATH THE ROOTS

Shallow cables may need to be buried deeper than normal.

6

BACKFILL CAREFULLY

Use sand as backfill around the roots. Hand tamp around and between the roots. Don't use a whacker, except on the very top layer of tarmac. If you're not sure what to do, ask the boss!

Figure 4.5. A guide to cable trenching along streets with trees produced by the Black Country Urban Forestry Unit, West Midlands, UK.

lead to some die-back. Youth plays its part: younger trees can stand more loss. Seedlings can (and do) lose as many as 90–98% of roots when transplanted as bare-rooted stock (i.e. with only the larger woody roots intact) and still survive. A simple rule of thumb is that when transplanting, a tree should have a 12 inch (30 cm) diameter root ball for each 1 inch (2.5 cm) of trunk diameter.

The odds of a transplanted tree surviving is increased if the nurseryman has

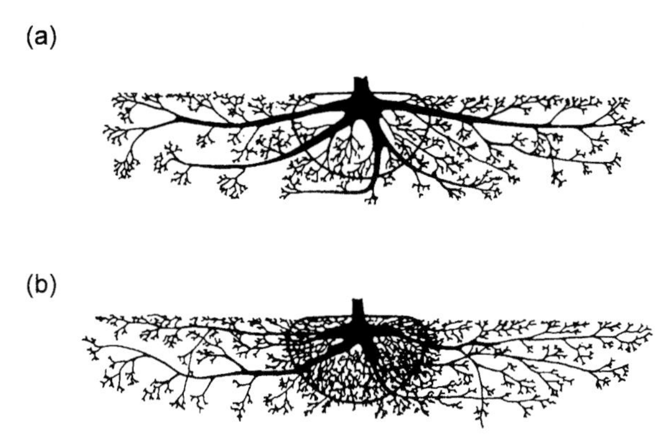

Figure 4.6. Root pruning and its effect on the production of new roots: (a) an unpruned tree with few roots in the root ball, and (b) a pruned tree with many fine roots in the root ball. From: Watson, G.W. (1986). Cultural practices can influence root development for better transplanting success. *Journal of Environmental Horticulture*, 4, 32–34.

encouraged a dense bushy root ball by root pruning: cutting through far-reaching roots a short way from the tree sometime before the tree is moved. Once a main woody root has been cut, new branches are produced just behind the damage over the next few months. A root 3–5 cm in diameter can produce up to 30 new tips (Figure 4.6). However good the transplanting, long-term survival is dependent upon growth of new fine roots. Temperate trees are best planted in the autumn and early winter to stimulate new fine-root development without a great demand for water from the shoot. Since evergreen trees are usually good at controlling water loss (see Chapter 2) they can tolerate spring planting.

How much soil is needed?

Trees planted in holes are often just as much in a pot as are trees planted into containers, either because of the compacted hole sides (see above) or because of surrounding buildings and roads. Just how much soil do we need to give them? A general rule of thumb in temperate areas (Europe and the USA) is that a tree requires about 1/2 – 3/4 cubic metres of soil per square metre crown projection (the shadow cast by an overhead sun). This translates into around 5 m^3 of soil for a medium-sized tree and up to 60–150 m^3 for a big tree. In effect, the smaller the amount of soil, the more likely that the tree will suffer drought. Such a tree will grow more slowly than an open-planted sibling because the roots are restricted (as discussed at the beginning of the chapter). But it will grow, and as the canopy gets bigger and needs more water, and the roots can't enlarge, it will grow closer and closer to the edge of drought and disaster.

Is the water table needed?

Trees will grow in restricted soil volumes but tend to pay the price of slow growth and early death. Moreover, as we saw in Chapter 2, trees can use prodigious quantities of water: hundreds of litres per day. These facts usually lead people to think that, for trees to grow really big, they need access to the abundant supplies of the water table. But they don't! Trees on the whole are dependent on rainfall for their water needs.

The empirical evidence can be seen in trees growing at the edge of tall cliffs, where the water table is completely out of reach. Moreover, trees will suffer in dry years just like any other plant, but seldom suffer if the water table is dropped as is all too common through borehole extraction of drinking water. Admittedly, trees do have access to deeper supplies of water than most herbaceous plants. In fact a number of trees and shrubs, both temperate and tropical, can absorb water with deep roots and pass it into the shallow roots and out into the soil. Emerman and Dawson (1996) in New York have shown that in a mature sugar maple this amounts to around 100 l per night. The mechanism appears to be that the roots act like wicks, helping water seep from wet to dry soil. The benefits are obvious; water is brought up during the night ready for use the next day. But this supply is still directly from rain rather than the water table. A more calculated analysis of the amount of water held in soil, with the same conclusion, is given in Box 4.2.

Having said all this, there are some trees that *do* rely on long roots reaching the water table (officially called phreatophytes). These include the tamarisk species (*Tamarix*) around the world, which grow in salty areas, and many trees of extreme arid areas.

Increasing water and nutrient uptake

There is a limit to how many roots a tree can produce because of the cost of producing and maintaining them. One way around this is to grow roots where they can absorb most.

Canopy roots

Rainforest trees (both tropical and temperate) are often festooned with thick mats of epiphytes (orchids, bromeliads, mosses, etc., which hitch a ride on the tree but take nothing from it), which are very good at extracting nutrients from the atmosphere and from water running down the host trunk. Some of the festooned trees produce 'canopy roots' from the trunk and branches to exploit the high nutrient content of the humus produced in these mats and caught in branch forks, hollows, etc. One tree of the Australian rainforest (*Ceratopetalum virchowii*) has canopy roots in clumps or enveloping the trunk like a fibrous coat but which

Box 4.2. Can a tree get enough water from the soil without using the water table?

Assume that a modest tree (15 m high with a canopy of 4 m radius) is using 40 000 litres of water per year. Over a growing season of 6 months/180 days this means 220 litres per day.

Root spread is likely to be something over 8 m in radius. Let's choose 8.4 m to keep the sums easy! A circle of 8.4 m radius, gives a rooting area of 220 m^2 (πr^2).

In a day the tree therefore needs 220 l from 220 m^2 of soil or 1 l from 1 m^2.

Now, 1 litre is 1000 cm^3 of water and 1 m^2 is 10 000 cm^2
So 1 l from 1 m^2 becomes:
1000 cm^3 from 10 000 cm^2
= 0.1 cm^3 from each cm^2 of soil surface.

This is equal to 0.1 cm (=1 mm) height of water from each square centimetre of soil each day during the growing season. This is the right order of magnitude; a whole forest uses 2–3 mm per day. Over 180 days the tree needs 180 mm of water.

The average rainfall in Britain is 600 mm per year. Not all of this will be available to the roots (some is intercepted and evaporated from the canopy, other plants will root in the same area, etc.) but there is more than enough to supply the tree.

A good loam can hold 130–195 mm of rain in 1 m depth. Sandy or gravelly soil may have nearer 50 mm stored. This is enough water for the entire growing season on loam or nearly two months on gravel if we can assume that this is all available to the tree. We undoubtedly can't assume that, but accepting these figures even within an order of magnitude does illustrate that trees do not normally need the water table.

do not reach down to the soil. These roots are able to recover nutrients from water running down the trunk that have been leached out higher up the canopy. The red mangrove (*Rhizophora mangle*) does the same sort of thing with sponges that grow on its prop roots: it grows a mass of fibrous roots into the live sponges. But in this case both partners benefit. The mangrove gets nitrogen from the sponges and they in return are given carbohydrate from the roots.

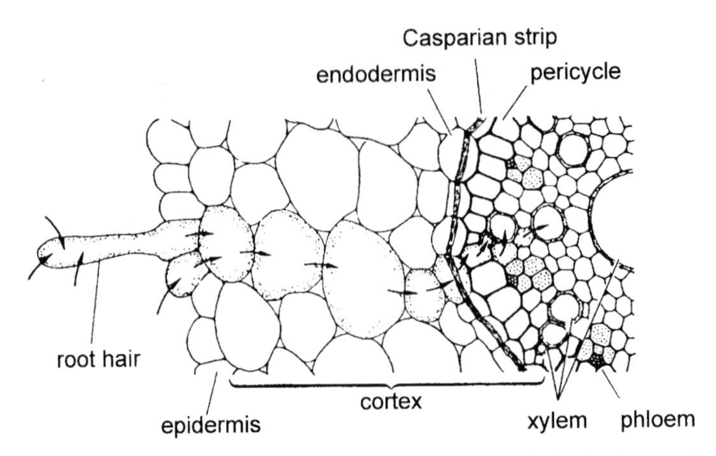

Figure 4.7. A root cross-section showing the passage of water and dissolved minerals from a root hair through the Casparian strip into the xylem. From: Esau, K. (1960). *Plant Anatomy*. Wiley, New York.

Trees, in common with other plants, have come up with several ways of increasing the surface area of the underground roots.

Root hairs

Root hairs are outgrowths of root cells behind the apex, which greatly increase the root's intimate contact with the soil (Figure 4.7). In red oak they are just 0.25 mm long but can vary from 0.1 mm in apple to 1.0 mm in blackcurrant. Most live for only a few hours, days or weeks, being replaced by new hairs as the growing tip elongates. Some trees, including honey locust (*Gleditsia triacanthos*) and the Kentucky coffee-tree (*Gymnocladus dioicus*), retain root hairs for months to years but such persistent root hairs appear relatively inefficient in absorption. On the other hand, despite their usefulness, root hairs are comparatively lacking in many trees and may be totally absent from some gymnosperms and at times from pecan (*Carya illinoensis*) and avocado (*Persea americana*). This is tied up with mycorrhizas.

Mycorrhizas

A mycorrhiza is an association between a root and a fungus which works to the benefit of both (symbiosis). Usually the tree gets extra water and hard-to-get nutrients (particularly phosphorus but also nitrogen and other nutrients), and perhaps some protection from fungal diseases and soil toxins such as heavy metal pollution. In return the fungus gets carbohydrates and other products from the tree. Many trees, including beech, oaks and pines, require the association to prosper; for others such as maples and birches it is not essential, while yet others,

Box 4.3. Types of mycorrhiza associated with trees

Ectotrophic mycorrhiza	Found in: - 90% of temperate trees of the northern hemisphere including most conifers - some southern hemisphere trees such as southern beeches (*Nothofagus*) and eucalypts (*Eucalyptus*) - less commonly in various families of the tropics including the dipterocarps (Dipterocarpaceae) The fungi concerned are usually Basidiomycetes, rarely Ascomycetes.
Endotrophic mycorrhiza, also called vesicular-arbuscular (VA) mycorrhiza	Found in a wide variety of families: - Aceraceae (maples) - Juglandaceae (walnuts and hickories) - Ulmaceae (elms) - Oleaceae (olives and ashes) - Magnoliaceae (magnolias and the tulip tree) - Hamamelidaceae (sweet gums amd witch hazels) - Cupressaceae (cypresses and junipers) - Araucariaceae (monkey puzzles and araucarias) - Ginkgoaceae (the ginkgo) - Taxodiaceae (redwoods and swamp cypress) The fungi are usually Zygomycetes.

notably members of the Proteaceae (one of the most prominent families of the Southern Hemisphere encompassing proteas, banksias, grevilleas and the macadamia nut), rarely, if ever, form mycorrhizas. Mycorrhizas are usually only found on trees that need them. Specimens on rich soil tend not to have mycorrhizas. This makes good economic sense; they cost the tree in carbohydrate and many fungi can become parasitic if the tree is too weak to provide the necessary sugars.

Mycorrhizas are found in around 80% of the world's vascular plants. Two types are found in trees. Ectomycorrhizas are found in only 3% of the flowering plants and are confined almost entirely to woody plants, primarily of the cool temperate regions of the northern hemisphere (Box 4.3). Endomycorrhizas are far more common in flowering plants and are found in a wide scattering of different trees. There are recorded instances where trees can form either type of mycorrhiza, sometimes both at once (e.g. poplars, willows, junipers, eucalypts and tulip trees, *Liriodendron* spp.). Within the same mycorrhizal type, trees may be infected with several species of fungi at the same time, and show a succession of different species as the tree ages.

Ectomycorrhizas appear on fine roots as a smooth fungal sheath coating side branches less than 0.5 cm long. Inside the sheath the fungus penetrates between the cells of the root to form a complex system of branching hyphae (the Hartig net) giving a large and intimate area for the exchange of materials. Outside the fungal sheath the fungus permeates out through the soil, often in well-organised strands, over large distances. The basic advantage to the tree is therefore to increase the effective size of the root system. Once infected, roots stop growing and lose their root hairs; they remain in one place and delegate exploration of the soil to the radiating hyphae. By contrast, fast-growing large-diameter woody roots on the same tree generally have root hairs and are non-mycorrhizal. Fungi forming ectomycorrhizas may only be able to invade slow-growing roots; perhaps the rarity of this type of mycorrhiza in the tropics may be due to continuous and rapid growth of fine roots, aided by the relative scarcity of basidiomycete fungi.

Endomycorrhizal roots look more normal. They keep their root hairs, and on the outside of the root the only usual sign of the fungus is a few short wispy hyphae. Most of the active part of the fungus grows into the cells of the root (not between cells as in ectomycorrhizas), producing highly branched structures (arbuscules) for the exchange of materials, and resting spores (vesicles); you may see this type of mycorrhiza referred to as VA or vesicular–arbuscular mycorrhiza.

The two sorts of mycorrhiza work in different ways although they provide comparable amounts of nutrients to their hosts. Litter from trees with ectomycorrhizas is harder and slower to decompose but the fungi can directly digest this litter and transfer nutrients straight to the tree. This is not cheap: 15% or more of the carbon fixed by the trees may be used by the large amounts of fungus involved. But on the cold, nutrient-limiting soils typical of ectomycorrhizal trees the benefits outweigh the cost.

By contrast, the rapidly decomposing litter of endomycorrhizal trees releases nutrients directly into the soil. The role of the fungus is to expand the volume of soil from which nutrients can be extracted. This may be just a matter of millimetres or centimetres from the root but enough to make a difference. The lesser role of the fungus is correspondingly cheaper for the tree: 2–15% of carbon fixed.

Endomycorrhizal fungi have low host specificity and completely unrelated plants may well be linked into a common mycorrhizal-mycelium system. This may help to reduce competitive dominance and promote coexistence and species diversity. It may also help explain why endomycorrhizal communities (e.g. tropical forest) are species-rich. Experiments have shown that phosphorus is preferentially transferred from dying roots of one plant to neighbours with the same mycorrhizal type. Ectomycorrhizal fungi tend to be more host-specific and are less likely to form widespread links but even here the joining of different individuals may be very important. For example, a study of Douglas fir seedlings (*Pseudotsuga menziesii*) in Oregon found all seedlings rooting into a hyphal mat produced by two species of ectomycorrhizal fungus which covered 28% of the forest floor. Since Douglas fir is shade-intolerant, the energy for survival when seedlings are in deep shade must come in large part from the mature trees. There is evidence that some pines (ectomycorrhizal) may in fact inhibit the establishment of endomycorrhizal trees and herbs. But it is not always that simple: under oak trees in Britain (ectomycorrhizal) the commonest seedlings can be ash and sycamore (both endomycorrhizal). There is still a lot to be discovered.

Other aids to nutrition

Trees in the pea family (Fabaceae; what used to be the Leguminosae), along with its herbaceous members, have swellings or nodules on the roots, which contain bacteria (one of several *Rhizobium* species) that can 'fix' atmospheric nitrogen into an organic form that is useable by the bacteria and the plant (Figure 4.8c). Again it is a two-way relationship; the tree gains nitrogen and the bacteria

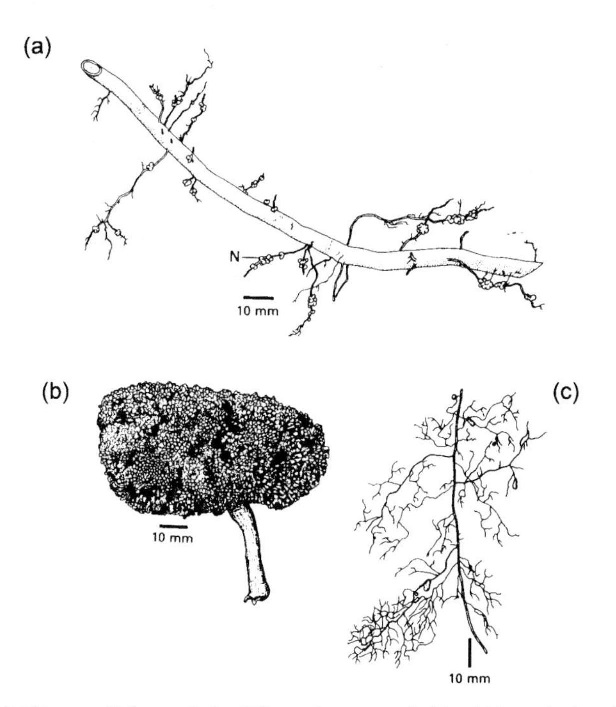

Figure 4.8. (a) Nitrogen-fixing nodules (N) on the roots of alder (*Alnus glutinosa*), (b) a single large nodule on alder, and (c) small nodules on the roots of an acacia (*Acacia pravassima*). From: Bell, A.D. (1998). *An Illustrated Guide to Flowering Plant Morphology.* Oxford University Press, Oxford. Reprinted by permission of Oxford University Press.

get a protected home and a supply of carbon. The benefit to the tree is exemplified by the false acacia (*Robinia pseudoacacia*), which is excellent at invading disturbed areas and is a fast-growing tree: up to 12 m in 10 years has been recorded. As in mycorrhizas, the formation of nodules and nitrogen fixation decreases as the level of soil nitrogen increases; fertilise the soil and the tree needs less from the bacteria.

Nodules can also be found outside the Fabaceae in a range of (primarily) temperate hardwood trees from a range of habitats: wetlands to deserts (Box 4.4). In these the nitrogen-fixing organisms are Actinomycetes (*Frankia* spp., related to filamentous bacteria). The nodules are not so much swellings as a series of short densely branched roots, giving a coral-like appearance. In alders (*Alnus* spp.) the nodules can reach the size of a tennis ball and live for up to a decade (Figure 4.8a, b). Perhaps most unusually, in the primitive Australian cycad *Macrozamia riedlei*, a photosynthetic bluegreen alga (an *Anabaena* species) invades the roots, which then grow up near or even above the soil surface, allowing the alga to photosynthesise and fix nitrogen.

Several nutritional aids can be used together. Red alder (*Alnus rubra*) has nodules and is commonly ectomycorrhizal, and its nitrogen fixation rates (up to 300 kg per hectare per year) are as high as published figures for the Fabaceae.

Roots are well known for their exudates released in response to nutrient deficiency, which are rich in organic and inorganic compounds. These influence the solubility of nutrients (by changing the soil pH) and their uptake by roots both directly and via their effects on soil microbes. This effect may only extend a few millimetres or centimetres into the soil (technically called the rhizosphere) but that's all that is needed. These chemicals can also help the plant to gain a competitive advantage, an effect called allelopathy, described in more detail in Chapter 9.

Root grafting

As mentioned when describing the root plate, the big roots produced by a tree near its base are commonly fused together into a solid network. What is equally interesting when thinking of tree nutrition is that roots of different trees can also graft together. In the relative stillness of the soil, roots readily intermingle without the same 'shyness' found between tree crowns (see Chapter 7). For example, in a mixed hardwood forest it is possible to get the roots of 4–7 trees below the same square metre of soil surface (Figure 4.9) with few areas where the woody roots of different species are more than 10–20 cm apart. Grafts between roots of different individuals of the same species are common in both conifers and hardwoods. Grafts between different species are possible but rare and many are 'false grafts' where the bark may fuse but the vascular tissue of the roots remain separate. The big framework roots are most likely to graft because they are the roots that grow most in diameter and so are most likely to press together. Indeed, it

Box 4.4. Trees that fix nitrogen in root nodules

Latin name	Common name	Family	No. of species nodulated /total no. of species in the genus	Notes
—	Leguminous trees and shrubs	Fabaceae	—	Including acacias, laburnums and pagoda tree; widespread
Coriaria	Coriarias	Coriariaceae	13/15	Warm temperate shrubs
Dryas	Avens	Rosaceae	3/4	Cool temperate dwarf shrubs
Purshia	—	Rosaceae	2/2	Western N America
Cercocarpus	—	Rosaceae	4/20	Western USA and Mexico
Casuarina	She-oaks	Casuarinaceae	24/25	Australia and Pacific coast
Myrica	Myrtles	Myricaceae	26/35	Almost cosmopolitan

Box 4.4. *(cont.)*

Latin name	Common name	Family	Nodulated spp./total spp.	Notes
Comptonia	Sweet fern	Myricaceae	1/1	N America
Alnus	Alders	Betulaceae	33/35	Mostly north temperate
Elaeagnus	Oleasters	Elaeagnaceae	16/45	Europe, Asia, N America and Australia
Hippophaë	Sea buckthorn	Elaeagnaceae	1/3	Europe and Asia
Shepherdia	Soapberries	Elaeagnaceae	3/3	N America
Ceanothus	Californian lilacs	Rhamnaceae	31/55	N America
Discaria	—	Rhamnaceae	2/10	Australasia and S America
Colletia	—	Rhamnaceae	1/17	S America

Based on: Torrey, J.G. (1978). Nitrogen fixation by actinomycete-nodulated angiosperms. *Bioscience*, **28**, 586–92. Becking, J.H. (1975). Root nodules in non-legumes. In *The Development and Function of Roots*, (ed. J.G. Torrey and D.T. Clarkson) pp. 507–66. Academic Press, London.

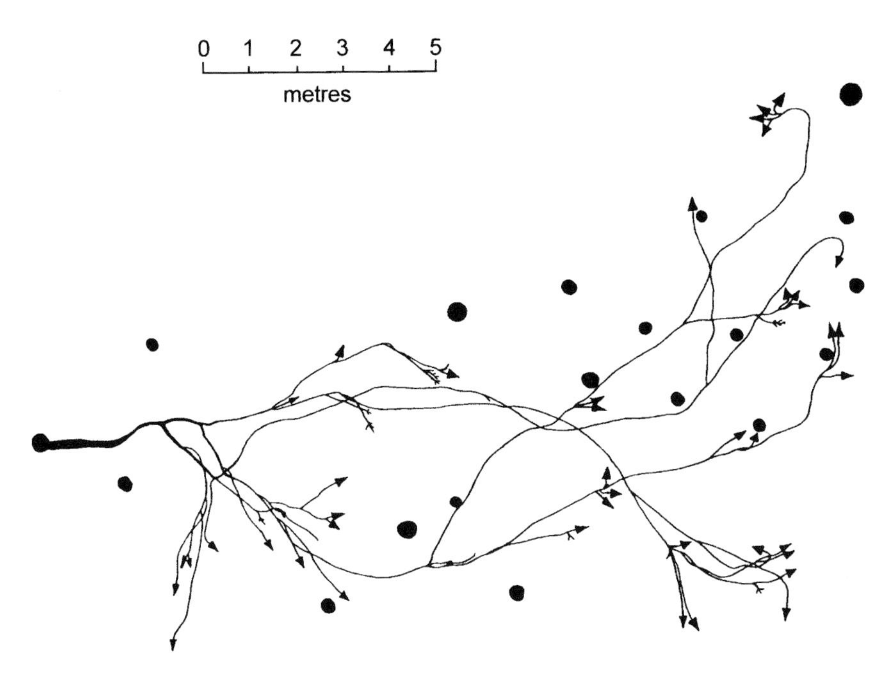

Figure 4.9. Plan view of a single lateral root of a red maple (*Acer rubrum*) about 60 years old. Circles show the location of other trees in the stand. Arrows indicate that the root tips were not found, these roots therefore continued somewhat further than is shown. From Lyford, W.H. and Wilson, B.F. (1964). Development of the root system of *Acer rubrum L. Harvard Forest Paper*, **10**.

is unlikely that fine non-woody roots ever touch despite their abundance, and thus they do not graft. (There are exceptions: the aerial roots of the strangler fig *Ficus globosa* have short hairs that will fuse together on contact, holding the bigger roots together, aiding grafting.) Since framework roots are usually involved, the frequency of root grafts between individuals depends on the closeness of trees. For example, in American studies almost all American elms (*Ulmus americana*) were root-grafted together if less than 2 m apart but this fell to 43% for elms up to 5 m apart and 29% for those up to 8 m apart.

Once grafted together there is considerable potential for transfer of materials between roots. Food and hormones flowing in the phloem move readily across grafts (although water and minerals in the xylem are less likely to exchange because they tend to flow mainly within the grain of one root). You may think that root grafting is another way for the big to become bigger at the expense of others, but it seems that the general movement is from the dominant individual to the underdog, although probably not enough to upset the dominance pattern. This is nicely illustrated when conifers are felled. Most conifers do not regrow once cut and the stump usually lives for no more than one growing season. But there are cases where stumps have been kept alive via root grafts for one or two decades, producing new rings and callus growth over the cut top without any leaves of their own. Root grafts between different species can have other bizarre consequences. The wood of a fig species that was root-grafted to an Indian tree, *Vateria indica*, that yields aromatic gum, had the smell of the *Vateria*, which in turn exuded a milky latex like that of the fig.

Root grafts between individuals are not always beneficial. Dutch elm disease and other diseases can be transmitted between trees via grafts. Herbicides applied to one tree can affect others by 'backflash' through root grafts. In this case, however, caution is needed; herbicides can move between trees not grafted together either by being exuded by roots and subsequently being taken up by other trees, or via mycorrhizal fungi and other organisms living around the root.

Food storage

Many herbaceous plants store food underground in modified stems and roots. This is a safe option because there are comparatively few large herbivores feeding underground; there are no underground cows. For a tree, however, the tough woody skeleton provides as much protection above ground as do woody roots below. Consequently, food in the form of starch is stored throughout the woody skeleton, above and below ground. In the event of the trunk being removed by, for example, fire or coppicing, the food stored below ground is used towards new growth. In extreme cases, this has resulted in the lignotuber, a large swollen root-collar in the soil covered in a mass of buds. Lignotubers are found in many tropical trees in seasonally dry savanna woodlands prone to frequent fires, but especially in the 'mallee' species of eucalyptus in W Australia. The largest lignotuber on record was 10 m across out of which branched 301 stems. The 'burs' of shrubby heathers, such as *Erica arborea* from which smoking pipes are made, are similar structures.

Development and growth of roots

Roots and shoots grow in a similar way (as described in Chapter 1): they get longer by new growth at the end (primary growth) and some also get fatter (secondary growth).

Root elongation: primary growth

As in herbaceous plants, a root lengthens from a small group of dividing cells (the meristem) just behind the root tip. The new growth is protected as it pushes through the soil by a root cap: a shield of tissue (which can be quite loose and easily wobbled with a fingernail) continually added to at the back and often lubricated by mucilage. Root caps are best developed in strongly growing roots and may be completely absent on short side roots. Behind the growing tip the new root differentiates into separate tissues. The centre of the root is dominated by the xylem (see Chapter 3), which radiates out into a number of star-like arms (technically called arches). In roots, and unlike stems, there is no central pith (Figure 3.2). Nestling in the gaps between the arms are strands of phloem. Around all this vascular tissue is a layer of cells (the pericycle) that gives rise to branch roots. This central 'stele' is surrounded by a one-cell thick cylinder (the endodermis), which in turn is surrounded by the cortex and outer skin (the epidermis). Water and dissolved minerals can pass relatively easily between the cells of the cortex, but in the endodermis the cells are tightly touching, and where they touch they are impregnated with suberin (a waterproof compound found in bark) to form the Casparian strip (Figure 4.7); in effect this strip ensures that all water and minerals that reach the internal plumbing have to pass through a cell and cannot sneak between cells.

You might ask how water is taken up by a root. At first thought it might seem that roots merely act like a sponge passively soaking up water. In actuality it is a more controlled process. Minerals in the water are actively 'pumped' through the Casparian strip and the water then follows passively by osmosis. This gives the root more control over what is taken up and by how much. For example, in some trees the tissue inside the Casparian strip is packed with sugars that have been stored over winter. These sugars draw in so much water that a positive 'root pressure' is built up, which forces water up the tree (just what this does for the tree is discussed in the previous chapter). Thus, the movement of water and minerals into the root is subject to the regulatory activity of the living cells. But as is discussed below, some water may get straight into the stele through cracks in old roots.

Side roots arise someway back from the root tip from the outer layer of the stele (the pericycle) usually opposite the points of the xylem star. These emerge through the parent root either by digesting a path or by sheer mechanical pressure. The new side root forms vascular connections with the stele of the main root and is then able to take part in the water uptake business.

Root thickening: secondary growth

For secondary growth to occur (i.e. thickening of the roots) there needs to be cambium present (again, see Chapter 3). This starts to develop before the end of the first year in the gaps between the xylem arms and soon joins together to form a complete cylinder between the xylem and phloem. Each year new xylem and phloem are produced (xylem on the inside and phloem on the outside) in annual rings, just as in the trunk, and the root gets thicker. As the roots thicken, a different type of cambium forms in the increasingly stretched pericycle: bark cambium. This produces corky cells, heavily laden with a waterproof waxy substance (suberin) to form bark similar to that above ground, complete with breathing holes (lenticels) (see Chapter 3). The cortex surrounding the root, increasingly cut off from the food supply of the stele, is starved, stretched and finally ruptured, to be shed with the endodermis (see Figure 3.2). Since the cortex often accounts for two thirds of the width of a new root, the young corky root may be somewhat thinner than the succulent tip. As a root grows only the growing tip remains unsuberised. In N Carolina, USA, both the loblolly pine (*Pinus taeda*) and the tulip tree (or yellow-poplar, *Liriodendron tulipifera*) were found by Kramer and Bullock (1966) to have usually less than 1% of the surface area of the roots unsuberised. It is analogous to a race: in fast-growing root systems large lengths may be unsuberised, but as growth slows down, suberisation catches up. Although the endodermis is shed, this does not necessarily mean free entry of water because the developing suberised bark has taken its place. There has been a long debate about whether suberised portions of root can absorb water (see Kozlowski *et al.* 1991) but the evidence points to substantial amounts of water getting into roots through holes and cracks in the bark such as at the ends of dead roots.

A cross-section of a root with secondary thickening increasingly resembles that of a branch. Upon looking closer, however, the wood of a root is commonly less dense than trunkwood (especially in hardwoods) with fewer fibres and a less clear distinction between growth rings. Moreover, ring-porous woods tend to become diffuse-porous with increasing distance from the trunk. In consequence, root wood often looks nothing like the wood in the trunk and requires specialised skill in its identification (see, for example, Cutler *et al.* 1987). Such structural differences appear to be due to the growing environment because roots of hardwoods exposed to light and air assume most of the characteristics of trunkwood.

Thickening in roots is concentrated at the base of the big lateral roots and in particular on their upper side, causing eccentric roots, which often bulge above ground, producing in extreme cases the buttress roots well known in tropical trees. Just why thickening should be so localised is still not completely understood. It may be because of the direct continuity of the upper side with the phloem in the trunk and better food supply. Also, however, portions of root exposed to air are known to grow better: they develop chlorophyll in the inner bark, which provides some food but more importantly leads to the production of growth hormones and local mobilisation of food reserves. It is also possible that the general thickening at the root collar may be due to stimulation of the cambium because it is under greater mechanical stress at this point (see Chapter 7).

Away from the trunk in the outer root system, secondary thickening is much less pronounced, resulting in the long rope-like roots that thicken slowly if at all. The production of wood is much more irregular than in stems and growth rings often vary greatly in thickness around and along a root, appearing in some parts and not others.

Speed of growth

Roots, like shoots, exhibit apical dominance; the main woody roots grow most rapidly and for longer periods compared with side roots. Small non-woody roots of red oak elongate at an average of 2–3 mm per day, whereas large-diameter lateral roots grow at 5–20 mm per day, and up to 50 and 56 mm per day in hybrid black poplar (*Populus* × *euramericana*) and false acacia (*Robinia pseudoacacia*), respectively. As noted above, mycorrhizal tips may not grow at all for the year or so they remain alive. At the other extreme are the delicate pink aerial roots of the liana *Cissus adnata* in the New World tropics, which may be no more than 1 mm thick but over 8 m long; they have been recorded as growing at more than 3 cm per hour after rain. The closely related *C. sicyoides* can grow at a constant 4 mm per hour day and night, a rate of almost a metre in a day! Most tree roots grow faster during the night.

Control of growth

Growth of the tops of trees, even tropical ones, is carefully regulated to be in tune with the changing seasons. Below ground, however, life is more anarchic, with roots growing whenever conditions are suitable. Temperature is most often the limiting factor. Roots normally grow best between 20 and 30 °C but in many trees continues down to 5 °C (this minimum may be as low as 0 °C for some northern trees or as high as 13 °C for citrus species). Thus in the tropics the roots of rubber trees (*Hevea* spp.) grow continuously in contrast to the rhythmic growth of their shoots. In temperate trees, roots may continue growing in winter long after the shoot system has become dormant, and in mild winters roots may grow right through to spring. The roots of red maple (*Acer rubrum*) in eastern N America are normally dormant between November and April but if they are kept warm at 20 °C the fine roots continue to grow at 5–10 mm per day even through the stem is dormant. As temperate soils warm in the spring, root growth (if it has stopped) usually resumes before the shoots. Lack of oxygen – for example by compaction of the soil or leaks from gas pipelines – can reduce or stop root growth but they will grow quite happily down to 10% oxygen (air contains 21%) and will only stop at around 3%. Water also probably has relatively little influence on root growth simply because roots are the last to feel water stress working down from the leaves, and are the first to recover. But temperate trees usually show a mid-summer lag as the soil dries to be followed by a new peak of root growth in late summer–autumn when shoots are growing very little. Root growth in hardwoods tends to peak in early summer whereas that of softwoods is more uniform through the season. This might be tied in with competition for resources within the tree: when the shoot is growing strongly it is better at attracting more than its share of food so slowing down root growth.

There has been a great debate as to whether root growth is governed purely by soil conditions or whether hormonal control, so evident in the shoot, extends down into the roots. On the one hand, root tips cut from a tree will grow quite happily in a moist dish, and roots of felled conifer stumps begin growing at the same time as those of neighbouring intact trees. On the other hand, roots still attached to the tree show cyclic periods of growth even when kept under uniform environmental conditions. On the whole, the shoot does have a definite influence on roots via supply of food and hormones. Part of the reason why this is not always obvious lies in the capability of the roots to store necessary food for use when the stem is dormant and in the fact that hormones in roots work at such low concentrations they are often difficult to detect. Root elongation is apparently controlled by one set of hormones (auxins) arising in the stem while the formation of new cells and lateral roots depends on a subtle balance between auxins and another set (cytokinins) produced by the root tips.

As mentioned before, root growth must inevitably be determined, at least in part, by the shoots to ensure an equitable balance between the amount of roots and shoots. Too few roots leads to lack of water in the canopy, whereas too many roots is a waste of resources that could be put towards canopy growth and reproduction. Fertilising a tree commonly leads to disproportionately greater increase in shoot growth. We normally think that by fertilising a tree it will grow big healthy leaves, make more sugar and so be able to grow more roots. But this assumes the root growth is limited by lack of carbohydrates and not by soil conditions or grass competition. Fertilisation may not always help.

Longevity of roots

In many ways the main structural roots are like the trunk and branches – long-lived – and the fine roots are like the leaves, flowers and fruits: regularly shed and regrown. The root system of a tree can be thought of as a persistent woody frame with disposable fine roots. Fine roots may live for just one week (as in some varieties of apple), one summer, or persist for 3–4 years (e.g. Norway spruce, *Picea abies*) growing year round (see above). In temperate zones, many small roots, especially those close to the surface, die in the winter; common walnut (*Juglans regia*) may lose more than 90% of its absorbing roots in winter, although others such as the tea plant (*Camellia sinensis*) may lose less than 10%. The shedding of fine roots is as ecologically important as above-ground leaf fall and the two are about equal in mass.

There is still uncertainty about what causes fine roots to die. Some people consider root death to be a normal physiological process governed by the tree's internal clock, just like leaf fall. Their arguments are based on the cost of keeping roots alive. For example, Reynolds (1975) puts forward the idea that when soil is dry (or 'physiologically dry' when soil and roots are too cold to work) the fine roots do nothing except consume food. He suggests that these roots would consume their own dry mass in reserves in one week. It therefore makes good economic sense to lose these wasteful roots and retain only a skeleton of woody roots from which new fine roots can grow when conditions improve.

Other people argue that root death is not due to the tree's actions but is a direct effect of unfavourable soil conditions and various pests and diseases. This fits better with the idea that roots are opportunists, growing even during the winter when conditions are suitable. Fine roots are very susceptible to dry conditions and may shrivel and die after only a few minutes of exposure to dry air (take note when planting trees!), and all remains can be gone within a few days. Nor have roots developed the extreme cold tolerance of shoots; temperatures below –4 to –7 °C can kill. Thus, the fine roots in woodland litter are easily exposed to lethal temperatures in temperate and northern climates. This explains why potted trees and shrubs (exposed to low temperatures from several sides) are less hardy and often need special protection in the winter. Fine roots can also be killed by movement of the root plate as the tree sways, and a whole range of subterranean predators from nematodes to small mammals are capable of damaging and cutting small roots (although little is known about their precise effect on trees). Fungal and insect attacks on the canopy can also strongly affect root mortality by reducing food and hormone export to the roots. In balsam fir (*Abies balsamifera*) attacked by spruce budworm, 'rootlet' death rose from less than 15% in healthy trees to more than 30% in trees with 70% defoliation, to more than 75% in completely defoliated trees.

Roots in wet soils

Oxygen in flooded soils is quickly used up by roots and microorganisms. Some oxygen diffuses down through the waterlogged soil but this is usually only enough to keep the top few centimetres of soil oxygenated. Below that the soil is devoid of oxygen (anaerobic). Woody roots can survive such conditions for some time when in a dormant state but fine roots die quickly, leading to poor water and mineral uptake, wilting of leaves and reduced photosynthesis. As well as the direct effect of oxygen starvation, there are problems of toxic compounds produced by the soil (e.g. hydrogen sulphide) and also directly by the tree. The roots of many *Prunus* species (including cherries, peach, apricot and almond, but not plum) contain cyanogenic glucosides, which break down to release cyanide gas when oxygen is limiting.

Trees vary tremendously in their ability to tolerate flooding. To some degree this is dependent on the state of the individual tree (older and dormant trees are generally more tolerant), but there is an underlying inherent difference between species (see Box 4.5). There are a number of trees such as the tupelo (*Nyssa aquatica*), swamp cypress (*Taxodium distichum*) and willows that can survive flooding for several months or even permanently. Others, such as alders (*Alnus* spp.) have *some* of their roots permanently in water. And these are not isolated examples.

Box 4.5. Flood tolerance of cultivated trees and shrubs exposed to a summer flood on the lower Fraser River Valley, British Columbia, Canada

Very susceptible to death and injury	Holly	*Ilex aquifolium*
	Hazel	*Corylus avellana*
	Lilac	*Syringa vulgaris*
	Mock orange	*Philadelphus gordonianus*
	Cotoneaster	*Cotoneaster* species
	Cherry	*Prunus* species
	Cherry laurel	*P. laurocerasus*
	Rowan	*Sorbus aucuparia*
	Japanese red cedar	*Cryptomeria japonica*
Less susceptible but often killed or severely injured	Hawthorn	*Crataegus laevigata*
	Box	*Buxus sempervirens*
	Blackberry	*Rubus procera*

Box 4.5. *(cont.)*

Some damage with leaf yellowing	Evergreen blackberry	*Rubus laciniatus*
	Pears	*Pyrus* species
	Roses	*Rosa* species
	Grapes	*Vitis* species
	Rhododendrons	*Rhododendron* species
	False acacia	*Robinia pseudoacacia*
No obvious injury	Apples	*Malus* species
	Walnuts	*Juglans* species
	Manitoba maple	*Acer negundo*
	Chestnuts	*Castanea* species

Based on Brink, V.C. (1954). Survival of plants under flood in the lower Fraser River Valley, British Columbia. *Ecology*, 35, 94–5.

The swamps and floodplains occupied by flood-tolerant trees can cover extensive areas; for example, some 2% of the Amazon basin (more than 100 000 km^2) is flooded by 2–3 m of water for 4–7 months every year. Saplings may be completely submerged for 7–10 months or longer each year. Yet these trees retain their leaves and photosynthesise even when completely submerged. How do such trees manage?

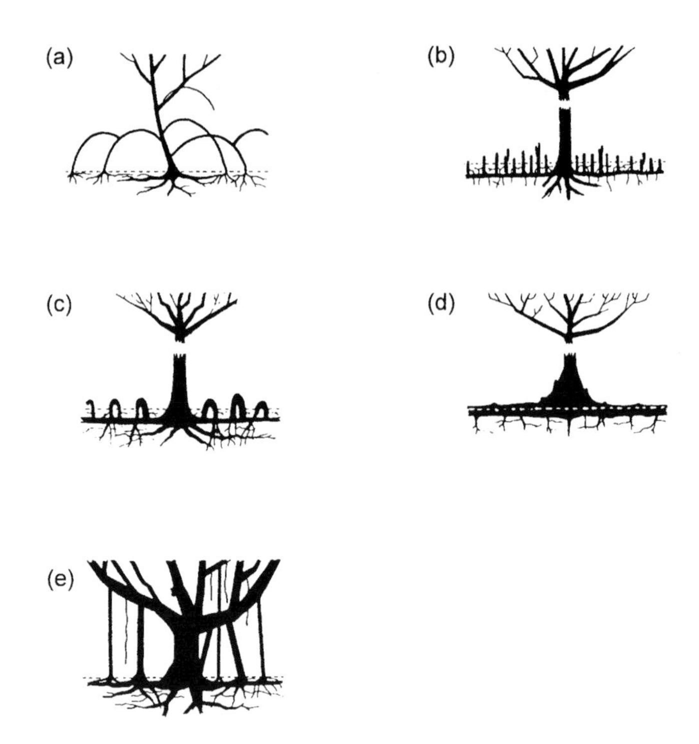

Figure 4.10. (a) Stilt roots and (b) peg roots of mangroves; (c) knee roots; (d) buttress roots on a tropical rainforest tree; and (e) pillar roots as seen on figs (e.g. *Ficus benjamina*). From: Jeník, J. (1978). Roots and root systems in tropical trees: morphologic and ecologic aspects. In *Tropical Trees as Living Systems* (ed. Tomlinson, P.B. and Zimmermann, M.H.) Cambridge University Press, New York.

What is visually most striking about many wetland trees, especially in the tropics, is the production of aerial roots (really 'adventitious' roots: roots appearing 'out of the usual place'), including stilt, peg and knee roots (Figure 4.10). Mangroves growing in tidal muds of the Old and New World tropics (e.g. *Rhizophora* species) produce a hula skirt of stilt roots from the trunk (often branched), which, once rooted, graft together into a rigid three-dimensional latticework. The mangrove may be held almost clear of the mud with the stilt roots taking the place of the sparse, short-lived normal roots. Other mangroves (e.g. *Avicennia* species) and several species in freshwater marshes (including the bay willow, *Salix pentandra*, in temperate areas, and various palms) have a shallow underground root system that produces short pencil-shaped peg roots, which stick out of the ground. These soft and spongy roots are rarely more than 1 cm in diameter but may be up to 2 m high and a single tree may produce as many as 10 000 of them. Both stilt and peg roots act as snorkels (and, certainly with stilt roots, provide stability in loose mud). Above ground the roots have large lenticels that feed into wide air passages (officially called aerenchyma) connected with the spongy air-filled underground roots. The stilt roots of *Rhizophora* have 5% gas space above ground, increasing to *c.* 50% after penetration into the mud. Oxygen is thus able to diffuse down into the underground root and toxic gases produced by incomplete aerobic metabolism can escape (although many flood-tolerant trees have biochemical adaptations to reduce the production, accumulation and impact of such toxins). The oxygen leaks out into the mud, forming an oxygenated envelope around the root, allowing it to function. This mechanism also works in the trunks of trees without modified roots (such as the tupelo, alders and ashes). The air flows down through the bark and air-filled sections of wood (xylem); the cambium was long thought to be impervious to gas, but it is now evident that this is not true of wetland trees.

Some flood-tolerant species produce new (adventitious) roots on the submerged part of the stem up near the water surface where there is more oxygen or even into the air. Such roots are found in a diverse range of trees from eucalyptus and bottlebrush (*Melaleuca*) species in the southern hemisphere to the tupelo, elms, willows and ashes of the north.

A number of unrelated tropical hardwood species, including mangroves (*Avicennia* spp.), and the important genus *Terminalia*, plus a palm or two in the genus *Phoenix*, can produce knee roots (properly termed pneumatophores) in wet areas. Here, a loop of root arches up into the air to form a distinct knee (Figure 4.10c). Again, knees act as aerating organs for the underground roots. The only conifer that can produce knees (and also the only non-tropical tree) is the bald or swamp cypress (*Taxodium distichum*) of seasonally flooded freshwater marshes in the southeast United States. These knees are produced in a different way, growing as a thickening of the upper surface of a horizontal root, producing the conical swelling that comes above the water. The knees are most frequent in regularly flooded sites and can grow to almost 4 m high, depending upon the normal annual high waters. However, whether they are actually involved in aerating submerged roots is uncertain; they may in fact be just for mechanical support, utilising the mass of roots that spreads from each knee as ballast (swamp cypress is very resistant to windfall).

Buttresses, pillars and stranglers

The aerial roots discussed above have been primarily on tropical wetland species, but they do also occur in temperate and dry-land trees (although they are not as common, as large or as variable). Stilt roots, for example, are well developed in the screw pines (*Pandanus* spp.) of the Old World tropics and a range of palms. These monocotyledonous trees do not have secondary thickening and therefore use the roots to hold up the increasingly tall and wide canopy. Other dry-land aerial roots are even more bizarre.

Buttresses

In temperate trees the root collar may flare a little way up the tree to form small buttresses. These flanges, which are really part root and part stem, are common in oaks, elms, limes and poplars, especially where the soil is not suitable for deep roots. For spectacular buttresses, however, we need to look to the tropics where each buttress can be many centimetres thick, extending out 1–2 m from the tree and rising 2–3 m, and sometimes 10 m, up the tree (Figure 4.10d). Some buttresses are thin and flat enough to be used to make walls of buildings; others are twisted and fused together to make a series of tank-like hollows. Buttresses are usually associated with tall trees that do not develop a tap root, and in some trees the stem all but disappears as it approaches ground level so that it is literally held up by the buttresses. That buttresses encourage mechanical stability seems beyond question; they help brace the roots to the trunk like angle brackets reducing the stress on the root collar when the tree sways. The big question is not what use they have but why they are rare outside the tropics. It may be that the large surface area makes them prone to damage by the temperature fluctuations of temperate climates or fire.

Pillar roots

The weeping fig (*Ficus benjamina*), the banyan (*F. benghalensis*) and a number of other figs use pillar roots beautifully. Slender free-hanging roots, which may be no more than 2 mm in diameter, grow down from the branches at up to 1 cm per day. Once anchored into the soil they form tension wood, which has the effect of contracting the root, so much so that they can lift large flower-pots from the ground. This contraction has the effect of straightening the root so that as the root thickens it forms a straight pillar able to bear the weight of the sideways spreading branches (Figure 4.10e). In this way, the trees can grow outwards to form groves whose 'trunks' are actually pillar roots. Indeed banyans may cover the largest area of any living plant; for example one tree planted in 1782 in the Royal Botanic Garden of Calcutta is 412 m (1350 ft) in diameter and covers an area of 1.2 ha (3 acres) with 1775 'trunks'.

Strangling figs

A variant of pillar roots is seen in the strangling figs. In the classic picture of the strangler fig, the seed germinates in the canopy of a tree and the roots rapidly grow down to the ground. Once there the roots rapidly thicken, branch and graft (there may also be an element of vertical contraction) to form a network of roots, which can be more than 30 m high, encasing the trunk of the host (Figure 4.11). Note that the 'trunk' of the strangler is really root. The strangler may then slowly kill the host by preventing it from increasing in girth: a slow version of the boa

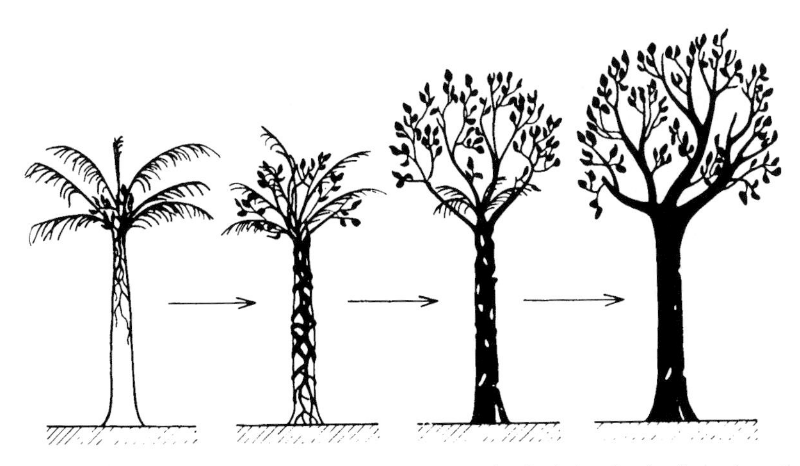

Figure 4.11. Four stages in the establishment of a strangler fig (*Ficus leprieuri*) on the palm *Elaeis guineensis*. From: Longman, K.A. and Jeník, J. (1987). *Tropical Forest and its Environment* (2nd edn). Longman Scientific and Technical, Essex.

constrictor! *Ficus leprieuri* has been seen to kill a large host tree in 30 years. The strangling part has probably been overdramatised and it is likely that competition with the host for water and light are probably more significant causes of host decline. Typically, however, less than 10% of infected hosts are killed, and many figs put down just a single harmless pillar root to ensure a water supply without having to grow their own trunk. Most stranglers (including the common rubber tree, *Ficus elastica*, and *F. benjamina*) will grow as normal but small trees if they germinate in the soil, as many keepers of house plants can testify. Nor are all stranglers figs. Strangling plants are also found in the unrelated genera of *Schefflera* (Araliaceae), *Clusia* (Clusiaceae), *Metrosideros* (Myrtaceae) and *Griselinia* (Cornaceae).

In temperate areas, the maidenhair tree (*Ginkgo biloba*) produces downward-growing woody knobs (called 'chichi': Japanese for nipples) from the underside of large old branches. These knobs, which look like aerial roots, bear buds and are in reality leafless branches, which grow down to the ground where they take root and produce new shoots. The largest ones measured, in Tokyo, were 2.2 m long and 30 cm in diameter.

Further reading

Brown, V.K. and Gange, A.C. (1990). Insect herbivory below ground. *Advances in Ecological Research*, **20**, 1–58.

Cutler, D.F. and Richardson, I.B.K. (1981). *Tree Roots and Buildings*. Construction Press (Longman), London.

Cutler, D.F., Rudall, P.J., Gasson, P.E. and Gale, R.M.O. (1987). *Root Identification Maual of Trees and Shrubs.* Chapman & Hall, London.

Ellison, A.M., Farnsworth, E.J. and Twilley, R.R. (1996). Facultative mutualism between red mangroves and root-fouling sponges in Belizean mangal. *Ecology*, 77, 2431–4.

Emerman, S.E. and Dawson, T.E. (1996). Hydraulic lift and its influence on the water content of the rhizosphere: an example from sugar maple, *Acer saccharum. Oecologia*, **108**, 273–8.

Gill, A.M. and Tomlinson, P.B. (1975). Aerial roots: an array of forms and functions. In: *The Development and Function of Roots* (ed. J.G. Torrey and D.T. Clarkson), pp. 237–60. Academic Press, London.

Gilman, E.F. (1988). Predicting root spread from trunk diameter and branch spread. *Journal of Arboriculture*, **14**, 85–9.

Griffiths, R.P., Castellano, M.A. and Caldwell, B.A. (1991). Hyphal mats formed by two ectomycorrhizal fungi and their association with Douglas-fir seedlings: a case study. *Plant and Soil*, **134**, 255–9.

Kelsey, P.J. (1987). Building movement and damage – identifying the causes. *Arboricultural Journal*, **11**, 345–61.

Khalil, A.A.M. and Grace, J. (1992). Acclimation to drought in *Acer pseudoplatanus* L. (sycamore) seedlings. *Journal of Experimental Botany*, **43**, 1591–602.

Kozlowski, T.T., Kramer, P.J. and Pallardy, S.G. (1991). *The Physiological Ecology of Woody Plants.* Academic Press, San Diego.

Kramer, P.J. and Bullock, H.C. (1966). Seasonal variations in the proportions of suberized and unsuberized roots of trees in relation to the absorption of water. *American Journal of Botany*, **53**, 200–4.

Lindsey, P. and Bassuk, N. (1991). Specifying soil volumes to meet the water needs of mature urban street trees and trees in containers. *Journal of Arboriculture*, **17**, 141–9.

Lyford, W.H. (1980). Development of the root system of northern red oak (*Quercus rubra* L.). *Harvard Forest Papers*, No. 21.

Nadkarni, N.M. (1981). Canopy roots: convergent evolution in rainforest nutrient cycles. *Science*, **214**, 1023–4.

Pulliam, W.M. (1992). Methane emissions from cypress knees in a southeastern flooplain swamp. *Oecologia*, **91**, 126–8.

Redmond, D.R. (1959). Mortality of rootlets in balsam fir defoliated by the spruce budworm. *Forest Science*, **5**, 64–9.

Reynolds, E.R.C. (1975). Tree rootlets and their distribution. In: *The Development and Function of Roots* (ed. J.G. Torrey and D.T. Clarkson), pp. 163–77. Academic Press, London.

Rolf, K. & Stål, Ö. (1994). Tree roots in sewer systems in Malmo, Sweden. *Journal of Arboriculture*, **20**, 329–35.

Rowe, R.N. and Catlin, P.B. (1971). Differential sensitivity to water logging and cyanogenesis by peach, apricot, and plum roots. *Journal of the American Society for Horticultural Science*, **96**, 305–8.

Torrey, J.G. (1978). Nitrogen fixation by actinomycete-nodulated angiosperms. *BioScience*, **28**, 586–92.

Wagar, J.A. and Barker, P.A. (1993). Effectiveness of three barrier materials for stopping regenerating roots of established trees. *Journal of Arboriculture*, 19, 332–9.
Watson, G.W. (1988). Organic mulch and grass competition influence tree root development. *Journal of Arboriculture*, 14, 200–3.
Wong, T.W., Good, J.E.G. and Denne, M.P. (1988). Tree root damage to pavements and kerbs in the city of Manchester. *Arboricultural Journal*, 12, 17–34.

Chapter 5: Towards the next generation: flowers, fruits and seeds

Like other plants, trees have to engage in sex by proxy, using the wind, water or an animal as an intermediary to get pollen from one tree to another (see Box 5.1). Unlike many other plants, the sheer size of trees raises extra problems of pollination, and eventually seed dispersal, which are solved in ingenious ways. The original trees, the conifers, were (and still are) wind-pollinated. The flowering plants (angiosperms), which include hardwood trees, evolved hand in hand with insects to be, not surprisingly, primarily insect-pollinated. Yet some have reverted back to the old way of wind pollination, and for very good reasons. These are linked to geography: most trees in high latitudes are wind-pollinated, but animal pollination (insects, birds and mammals) becomes more important the closer one gets to the tropics, reaching 95% of trees in the tropics. Figure 5.1 gives an overview of general flower structure.

Animal pollination

Animal pollination is primarily the world of the insect; in the wettest Costa Rican forests, for example, 90% of trees are insect-pollinated. But within insect pollination there are different strategies. Some trees, like magnolias, apples, rowan (*Sorbus aucuparia*), European spindle (*Euonymus europaea*), some maples, hawthorns (*Crataegus* spp.) and a long list of others, go for quantity. They are generalists that spread the pollen on a wide range of flies and beetles in the hope that some will arrive on another flower of the right species. Common features are open flowers facing upwards, a drab colour, many stamens, easily reached nectar and a strong scent especially at night (see Box 5.2 and Figure 5.2).

Other trees opt for quality, catering to a more limited number of specific pollinators that are more likely to go straight to another tree of the same kind. Bees, butterflies and moths fall into this category. Darwin noticed that certain pollinators are attracted to flowers of certain colours. Box 5.2 shows that each animal group tends to be attracted to different types of flowers. Notice that day-flying moths and butterflies, and bees, can be attracted to similarly coloured flowers: butterflies will visit predominantly bee-flowers such as privet (*Ligustrum* spp.), alder buckthorn (*Frangula alnus*) and lime (*Tilia* spp.) in Europe.

Colour can also be used to discourage insects from visiting flowers that are already pollinated. For example, in the horse chestnut (*Aesculus hippocastanum*), the markings inside the flowers change from yellow to red in pollinated flowers.

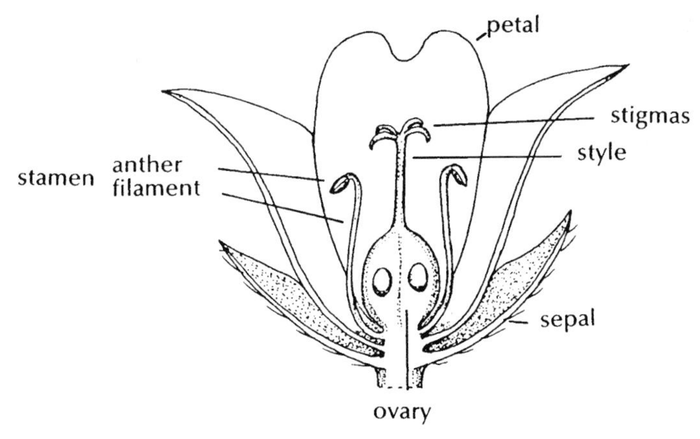

Figure 5.1. The make-up of a flower. The petals collectively form the corolla. The sepals collectively form the calyx. Together they form the perianth. The male stamens contain the pollen-producing anthers. The central ovary (containing the ovules, which will become the seeds) has a stigma(s) for catching pollen mounted on a style. From: Hayward, J. (1987). *A New Key to Wild Flowers.* Cambridge University Press.

Figure 5.2. Magnolia flower (*Magnolia wilsonii*), which appeals to a wide range of insects by offering an open flower facing upwards, a drab colour and many stamens.

At the same time, nectar production is reduced. To bees the red appears black and these unattractive flowers are avoided. The tree and the bee gain in the same way; precious time is not wasted on flowers that do not need pollen and which contain no reward. Why doesn't the tree simply shed the petals from the pollinated flowers? In experiments with Brazilian *Lantana* shrubs, which repel butterflies in a similar way, the insects homed in even on flower-heads with a large proportion of individually unattractive purple flowers. It seems the pollinated flowers of *Lantana* and horse chestnut are retained because they are bright and

Box 5.1. Flower and fruit characteristics for native and common introduced trees and shrubs of the British Isles. The flower types are explained in the text.

Common name	Latin name	Flowering date (months)	Pollinator	Flower type	Fruit/cone type
Hardwoods					
Alder	*Alnus glutinosa*	2–3	Wind	Monoecious	Woody cone containing nuts/ samaras
Alder buckthorn	*Frangula alnus*	5–6 (–9)	Insects (esp. bees)	Hermaphrodite	Drupe (2–3 stones)
Ash	*Fraxinus excelsior*	4–5	Wind	Dioecious or mixed	Samara
Beech	*Fagus sylvatica*	4–5	Wind	Monoecious	Nut

Berberis	*Berberis vulgaris*	6–7	Insects	Hermaphrodite	Berry
Birches	*Betula* spp.	4–5	Wind	Monoecious	Samara
Blackthorn	*Prunus spinosa*	3–5	Insects	Hermaphrodite	Drupe
Broom	*Cytisus scoparius*	5–6	insects (large bees)	Hermaphrodite	Exploding legume plus seeds with elaiosomes
Butterfly bush	*Buddleia davidii*	6–10	Insects (butterflies)	Hermaphrodite	Capsule
Bog myrtle	*Myrica gale*	4–5	Wind	Dioecious (but can change sex)	Nut with wings
Box	*Buxus sempervirens*	4–5	Insects (bees and flies)	Monoecious	Capsule and seeds with elaiosomes

Box 5.1. *(cont.)*

Common name	Latin name	Flowering date (months)	Pollinator	Flower type	Fruit/cone type
Buckthorn	*Rhamnus catharticus*	5–6	Insects	Dioecious, (rarely Hermaphrodite)	Drupe (3–4 stones)
Cherries	*Prunus* spp.	4–5	Insects	Hermaphrodite	Drupe
Cotoneaster	*Cotoneaster integerrimus*	4–6	Insects (especially wasps)	Hermaphrodite	Drupe (2–5 stones)
Crab apple	*Malus sylvestris*	5	Insects	Hermaphrodite	Pome
Currants	*Ribes* spp.	3–5	Insects	Hermaphrodite (rarely Dioecious)	Berry

Dogwood	*Cornus sanguinea*	6–7	Insects	Hermaphrodite	Drupe
Elder	*Sambucus nigra*	6–7	Insects (esp. small flies)	Hermaphrodite	Drupe
Elms	*Ulmus* spp.	2–3	Wind	Hermaphrodite	Samara
Gorses	*Ulex* spp.	3–6 or 7–9	Insects	Hermaphrodite	Exploding legumes and seeds with elaiosomes
Guelder rose/ wayfaring tree	*Vibernum* spp.	6–7	Insects	Hermaphrodite	Drupe
Hawthorns	*Crataegus* spp.	5–6	Insects	Hermaphrodite	Drupe (1 or more stones)
Hazel	*Corylus avellana*	1–4	Wind	Monoecious	Nut

Box 5.1. *(cont.)*

Common name	Latin name	Flowering date (months)	Pollinator	Flower type	Fruit/cone type
Holly	*Ilex aquifolium*	5–8	Insects (honey bees)	Dioecious, rarely Hermaphrodite	Drupe (3+ stones)
Hornbeam	*Carpinus betulus*	4–5	Wind	Monoecious	Nut and bract
Horse chestnut	*Aesculus hippocastanum*	5–6	Insects (bees)	Hermaphrodite (with some male flowers)	Capsule
Laburnum	*Laburnum anagyroides*	5–6	Insects (bumble bees)	Hermaphrodite	Exploding legumes and seeds with elaiosomes
Limes	*Tilia* spp.	6–7	Insects (bees)	Hermaphrodite	Nuts with bracts

Maples	*Acer* spp.	4–7	Wind and small insects	Hermaphrodite or mixed	Samara
Oaks	*Quercus* spp.	4–5	Wind	Monoecious	Nut
Pear	*Pyras pyraster*	4–5	Insects	Hermaphrodite	Pome
Poplars	*Populus* spp.	2–4	Wind	Dioecious, some rarely Monoecious	Capsule with plumed seeds
Privet	*Ligustrum vulgare*	6–7	Insects	Hermaphrodite	Berry
Rhododendrum	*Rhododendrum ponticum*	5–6	Insects	Hermaphrodite	Capsule with small winged seeds
Roses	*Rosa* spp.	5–7	Insects	Hermaphrodite	Achenes enclosed in a fleshy hip

Box 5.1. *(cont.)*

Common name	Latin name	Flowering date (months)	Pollinator	Flower type	Fruit/cone type
Sea-buckthorn	*Hippophaë rhamnoides*	3–4	Wind	Hermaphrodite sometimes Monoecious	Achene with fleshy receptacle
Snowberry	*Symphoricarpos albus*	7–10	Insects (bees and wasps	Hermaphrodite	Drupe (2 stones)
Spindle	*Euonymus europaeus*	5–6	Small insects	Hermaphrodite or mixed	Capsule and seeds with arils
Spurge laurel/ mezereon	*Daphne* spp.	2–4	Insects (moths and bees)	Hermaphrodite	Drupe
Strawberry tree	*Arbutus unedo*	9–12	Insects (?)	Hermaphrodite	Warty berry

Sweet chestnut	*Castanea sativa*	7	Insects	Monoecious	Nut
Whitebeams	*Sorbus* spp.	5–6	Insects	Hermaphrodite	Pome
Willows	*Salix* spp.	2–5	Insects and birds	Dioecious, some rarely Monoecious	Capsule with plumed seeds
Conifers					
Juniper	*Juniper communis*	5–6	Wind	Dioecious	Fleshy cone
Scots pine	*Pinus sylvestris*	5–6	Wind	Monoecious	Cone
Yew	*Taxus baccata*	3–4	Wind	Dioecious	Seed with aril

Based on information from Clapham, A.R., Tutin, T.G., and Moore, D.M. (1987). *Flora of the British Isles.* Cambridge University Press; Snow, B. and and Snow, D. (1988). *Birds and Berries.* Poyser, Calton, Staffordshire; and Sedgley, M. and Griffon, A.R. (1988). *Sexual Reproduction of Tree Crops.* Academic Press, London.

Box 5.2. Types of flowers associated with different animal pollinators

Pollinator type	Flower			
	Type	Colour	Smell	Reward
Beetles	Upwards facing bowl	Brown, white	Strong	Pollen, nectar
Flies	Upwards facing bowl	Pale, dull	Little	Nectar
Bees	Often asymmetrical, strong, semi-closed	Yellow, blue	Fairly strong	Nectar
Butterflies, moths	Horizontal or hanging	Red, yellow, blue (day); white or faintly coloured (night)	Heavy, sweet	Nectar
Birds	Hanging or tubular, copious nectar	Vivid red	Absent	Nectar
Bats	Large strong single flowers, or brush-like	Greenish, cream, purple	Strong at night	Nectar, pollen

Figure 5.3. Flower of a eucalypt (*Eucalyptus stellulata*). From: Williams, J. and Woinarski, J. (1997). *Eucalypt Ecology*. Cambridge University Press.

conspicuous and attract insects from a distance; the flower colour directs the insect to individual flowers once it has arrived. In a similar way, woody plants, such as dogwoods (*Cornus* spp.) and bougainvillea (*Bougainvillea spectabilis*), use large petal-like bracts (really modified leaves) to help make small, inconspicuous flowers more attractive to pollinators; and the European guelder rose (*Viburnum opulus*) uses large sterile white flowers around the outside of the flower head to advertise the smaller and less showy fertile flowers in the middle.

Birds are also important pollinators of trees, especially in the tropics with hummingbirds and honeycreepers in the New World and honeyeaters and lorikeets in the Old World. Bird-pollinated flowers tend to conform to a pattern (Box 5.2), exemplified by the fuschias commonly grown in gardens: tubular flowers (to hide the copious nectar from the wrong pollinators), brightly coloured orange or red (although many Australian bird flowers such as eucalypts are yellow or white) but odourless (birds have a poor sense of smell). However, in the dry climate of Australia the bottle-brush trees (*Banksia* spp.) and other related members of the Proteaceae are typical bird-pollinated flowers with masses of protruding stamens to dust the birds' feathers as they suck the copious nectar. The eucalypts take this further; the petals have been lost and the calyx is modified into a cap over the flower (the operculum), which is pushed off as the bud opens; the great tuft of stamens is the real attractant to birds (Figure 5.3).

Although commonest around the tropics, bird pollination is occasionally found in temperate areas. In Europe small warblers and blue tits have been seen taking nectar from gooseberry, cherry and almond. A most striking example are the furry catkins of willows which for many years were thought to be wind- and insect-pollinated. Each flower in a catkin has a large nectary producing a large and easily seen glistening drop of nectar. There are many records of blue tits feeding on this nectar, which is readily accessible to such a small beaked bird (see Kay 1985 for further details). Pollen is clearly seen liberally dusting the face and chest feathers as they fly from bush to bush. Despite being large and warm-blooded, and so needing a lot of energy, it is calculated that blue tits in Britain can get their total energy needs for a day in less than four hours of feeding on willows (although they may need to balance their diet with insects). This demonstrates how rich and abundant nectar can be in bird-pollinated flowers. Indeed, the nectar from Australian bottle-brush trees is gathered for food by the aboriginals. Bird pollination is thus very expensive to the plant; is it worth it? Where there is a comparative dearth of highly developed flower-visiting insects, such as in the tropics, or where the tree is hedging its bets, as in the willow, the answer appears to be yes. Certainly birds are very good pollinators and will move pollen further than bees. They also show considerable constancy to flowers of a single species and can visit many thousands of flowers a day.

Mammals also make good pollinators, especially the flying ones. Bats pollinate a number of tropical trees including the kapok or silk cotton tree (*Ceiba pentandra*), the fruit of which has been used to stuff furnishings; balsa (*Ochroma lagopus*) and the infamous durian (*Durio zibethinus*) whose fruits smell like sewers and old socks but are considered a delicacy by the people of southeast Asia! Bat-flowers tend to be large dull things (bats are colour-blind) which open at night, producing copious nectar and a sour or musty smell. Many species flower when they are leafless, while others produce flowers in the open on the trunk or edge of the canopy to give plenty of room for manoeuvre. Some bats take nectar while hovering; others land (requiring a strong flower to take the weight). The baobab (*Adansonia digitata*) is an example of the latter (Figure 5.4); the bat hangs onto the tufts of purple stamens, from which they lap up the nectar.

Non-flying mammals are poor cross-pollinators. Nevertheless, a variety of monkeys and possums are implicated in pollination of trees. Some proteas (*Protea* spp.) of South Africa are pollinated by rodents, and the traveller's palm in Madagascar (*Ravenala madagascariensis*) by a lemur. In Australia the bird-pollinated banksias and eucalypts are also visited by marsupial mice and honey possums, equipped with long tongues for feeding on nectar, and fur that picks up pollen. Perhaps most bizarre is the knobthorn acacia (*Acacia nigrescens*) in semi-arid savannas of Africa that appears to be pollinated by giraffes, which come to eat the flowers at the end of the dry season! Although the trees lose flowers (they make up to 40% of the giraffes' annual diet) the long-legged animals can travel over ten miles a day between stands of trees, ensuring pollen is well spread.

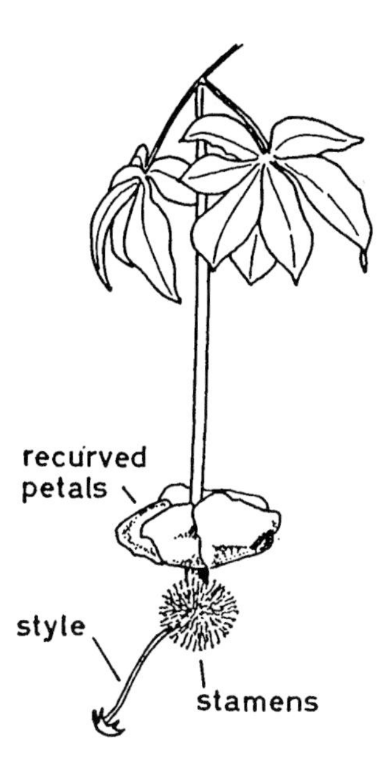

Figure 5.4. Bat-pollinated flower of the baobab tree (*Adansonia digitata*). From: Proctor, M.C.F. and Yeo, P. (1973). *The Pollination of Flowers.* Collins, London. With permission from HarperCollins Publishers Ltd.

Treating your guests right

Flowers are not always passive dining halls: they can be quite abusive to their guests. Broom (*Cytisus scoparius*), a European shrub of heaths and open woods, has explosive pea-like flowers. As the bee lands, the two bottom 'keel' petals are pushed apart by the insect's weight, releasing the sprung stamens and style. Five short stamens hit the bee's underside and five long ones and the style strike the bee on the back of the abdomen (Figure 5.5). Like a firework, each flower is used just once, delivering and collecting pollen in one explosion to an insect of the correct weight. The mountain laurels (*Kalmia* spp.) of N America are also explosive. Here the ten stamens arch back with the tip of each held in a little cavity in the petal (Figure 5.6). When an insect lands on the flat flower and pushes the petals down, the stamens are released with a jerk, dusting the unsuspecting insect. Other shrubs have stamens that are not captive but 'irritable'. The closely related barberrys (*Berberis* spp.) and Oregon grapes (*Mahonia* spp.) have six stamens pressed against the petals. When an insect pushes against the base of the stamens they spring inwards within 45 thousandths of a second to dust the insects with pollen. Unlike the one-off mechanisms in broom and mountain laurels, the barberry stamens slowly bend back to their original position over several minutes, although *Mahonia* only returns to an erect position.

Attracting the right pollinator is a matter of providing the right attractions and the right rewards. Nectar is mostly sugars with (usually) small quantities of amino

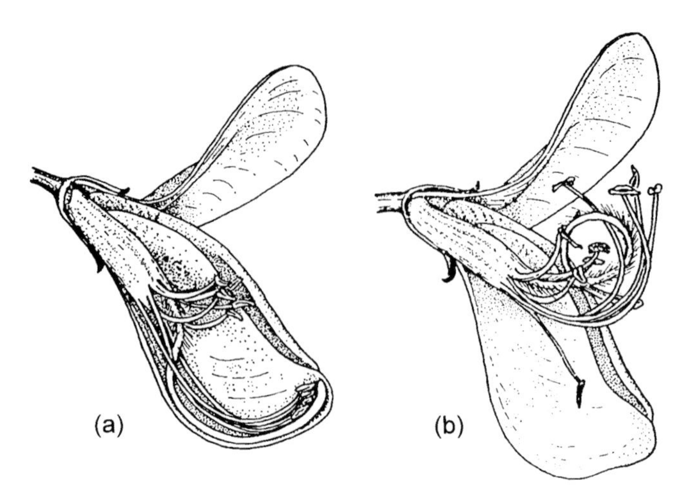

Figure 5.5. Broom (*Cytisus scoparius*). (a) A newly opened flower with half the calyx and corolla removed to show the stamens and style held under tension. (b) The flower 'exploded' following an insect visit. From: Proctor, M.C.F. and Yeo, P. (1973). *The Pollination of Flowers*. Collins, London. With permission from HarperCollins Publishers Ltd.

Figure 5.6. Flowers of sheep laurel (*Kalmia angustifolia*) from eastern Canada, showing the captive stamens.

acids to supplement the diet of those animals that largely depend upon it for food. The precise nectar composition is tailored to the preferences of the main pollinator. Pollen is also used as a food by many animals, including bees, some butterflies and bats. This may also be tailored to the pollinator; in some New World trees, at least, the pollen contains amino acids that are useless to the plant but essential for its bat pollinators.

Trees may provide other rewards. The cocoa tree (*Theobroma cacao*)—from which we derive chocolate—is pollinated by biting midges that breed in the decaying pods. Other tropical trees 'pay' moths one leaf crop which is devoured by the caterpillars; in return the adults do the pollinating. Perhaps the ultimate relationship between plant and pollinator is found in the figs, many of which are woody trees or stranglers (Chapter 4). Here the 'flower' looks more like a gourd-like green fruit because the base of the flower-head has grown up around the flowers to create a hollow sphere lined on the inside with male and female flowers (a syconium, Figure 5.7i). The small female fig wasp, which does the pollination, has to squeeze in through a small hole, losing her wings and antennae as she does so (Figure 5.7ii). Once inside, the mutilated female wasp lays her several hundred eggs into the flowers (oviposition). At this stage the female flowers are ripe and are pollinated as she wanders around with the pollen she has brought with her. The eggs hatch and the grubs feed and develop inside the flowers, their presence stimulating the tree not to drop that fruit (Figure 5.7iii). Not all is lost for the fig because it has two layers of ovaries; the upper are infested with grubs but the lower are too deep to have received eggs and quietly get on with growing seeds. Several weeks later (Figure 5.7iv) tiny wingless males hatch out, locate female-containing flowers, chew their way in and mate. Their last job is to tunnel out of the fig, but since they are wingless they are doomed to go nowhere. Once the fig is punctured, the accumulated high level of carbon dioxide leaks out. This is the signal for the male flowers to ripen and for the females to emerge. On their way out of the exit holes made by the male flies, the female wasps collect pollen. With some wasps this is a matter of getting coated on the way out, but in others the females quite literally stop and stuff special pockets with pollen using their legs. Then the cycle starts again with the female hunting by smell for another flower of the same species, leaving the fig of its birth to ripen, change colour and hopefully be eaten to spread the seeds. In case this puts you off eating figs ever again, relax, figs for consumption are parthenocarpic (see below), developing without seeds or fertilisation!

Keeping the wrong animals away is a matter of hiding away the pollen and nectar so only the right insects will get the reward. An obvious example is putting nectar at the end of a long spur so that only long-tongued butterflies or moths will bother visiting. Under this category can be classed 'buzz pollination'. Some tropical trees, exemplified by the Indian laburnums (*Cassia* spp.) in the New and Old World (and the tomato, *Lycopersicon esculentum*!) have anthers that do not split open to reveal the pollen, as is normal, but have small openings in the end of the anther. The only way to get the pollen out is to shake the anther at the right frequency. An alighting bee sits over the stamen and rapidly vibrates its flight muscles (not the wings), causing the pollen to spray out of the pores and catch on the hairy bee. This is quite an event and can be heard up to 5 m away.

Pollinators, of course, are not dumb servants and have ways of cheating the system. *Vespula* wasps in Europe, for example, steal nectar from heathers by boring holes in the sides of the flowers. Numerous similar tropical examples involv-

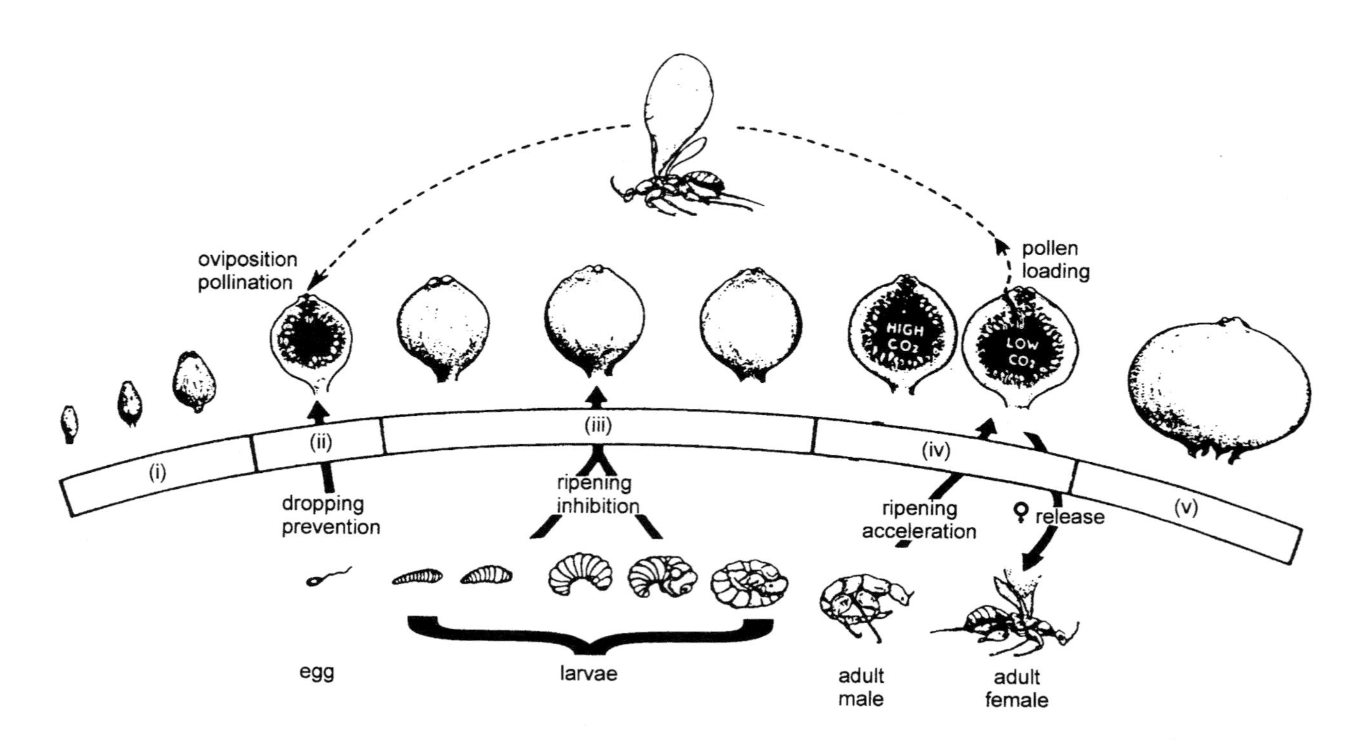

Figure 5.7. Fig pollination. Modified from: Galil, J. (1975). Fig biology. *Endeavour, New Series*, 1, 52–6. With permission from Elsevier Science.

ing insects and birds could have been chosen. This entrepreneurial spirit explains why exotic trees not known for self-pollination can sometimes set seed even when their natural pollinator is absent. But this is not always so: oil production from the oil palm (*Elais guineensis*), originally from W Africa, was greatly improved in Malaysia after the introduction of its natural pollinator, a weevil.

Wind pollination

Wind pollination is often seen as being primitive and wasteful in costly pollen and yet it is surprisingly common, especially in higher latitudes. There must be a good reason for this. It is tempting at first to say that this is explained by there being fewer insects and higher wind speeds in northern forests. But the whole answer is a little more subtle. Wind is very good at moving pollen a long way; pollen can be blown for hundreds of kilometres, only birds can get pollen anywhere near as far. The drawback is that wind is obviously unspecific as to where it takes the pollen. It is like trying to get a letter to a friend at the other end of the village by climbing onto the roof and throwing an armful of letters into the air and hoping that one will end up in their garden. For the relatively few dominant tree species that make up temperate forests, where there are many individuals of the same species within pollen range, this is quite a safe gamble. If all my friends around me were throwing letters off roofs, I'd be bound to get one. Indeed, wind pollination is found in other groups with large frequent clones such as grasses, sedges and rushes. By contrast, in the tropics where each tree species has few, widely scattered individuals, the chance of wind blowing pollen to another individual is sufficiently slim that animals are a safer bet. Even tall emergent trees in the tropics are usually not wind-pollinated despite being in windy conditions. In a similar way, trees in temperate forests that *are* insect-pollinated (such as whitebeams (*Sorbus* spp.), hawthorns and apples) tend to grow as solitary, widely spread individuals.

Modifying the flower for wind

Since wind-pollinated flowers have no need to attract insects or other animals, they have dispensed with bright petals, nectar and scent. These are at best a waste and at worst an impediment to the transfer of pollen in the air. The result is insignificant-looking flowers and catkins.

Despite looking so nondescript, wind-pollinated flowers and catkins are based on normal flower structure. The odd-looking mass of red elm flowers (*Ulmus* spp.) that appear in early spring before the leaves (Figure 5.8) are perfect little flowers with four stamens and a central ovary with two styles, all stretched out into the wind. The petals have been reduced to a four- or five-lobed fringe around the base. In many ways, oaks are similar except here the male and female parts are borne in separate flowers (the arrangement of sexes is further discussed

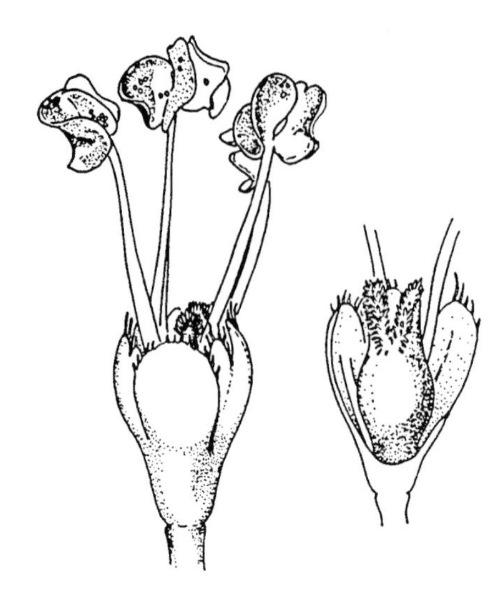

Figure 5.8. One elm flower from a mass, showing the four stamens and the ovary with stigmas in the cut-away section. From: Proctor, M.C.F. and Yeo, P. (1973). *The Pollination of Flowers.* Collins, London. With permission of HarperCollins Publishers Ltd.

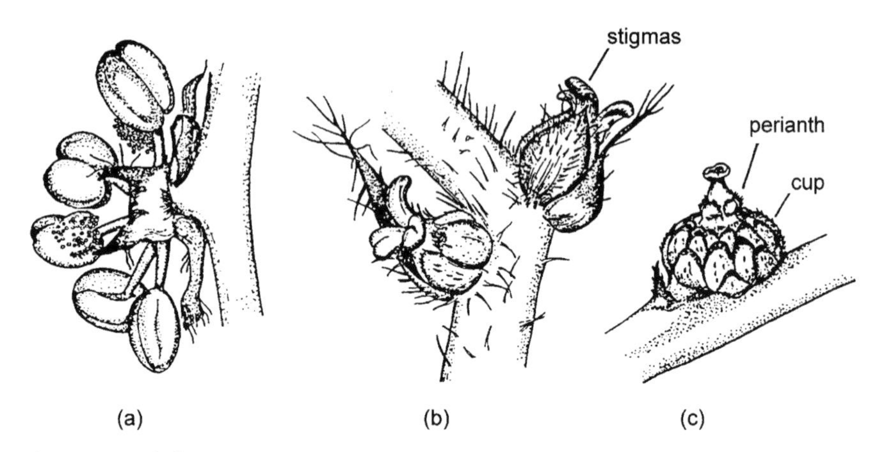

Figure 5.9. Oak flowers (*Quercus robur*). (a) side view of a single male flower on a catkin; (b) two female flowers; and (c) a young developing acorn. From: Proctor, M.C.F. and Yeo, P. (1973). *The Pollination of Flowers.* Collins, London. With permission of HarperCollins Publishers Ltd.

below) both appearing with the opening leaves. The male flowers are very simple; the remains of petals (a 4–7 lobed perianth; see Figure 5.1) are dwarfed by the 4–12 stamens offering copious pollen straight to the wind (Figure 5.9). To make them even more accessible to the wind the male flowers are loosely grouped together on hanging catkins up to 10 cm long. The female flowers are surrounded at their base by a series of overlapping bracts (modified leaves), which will go

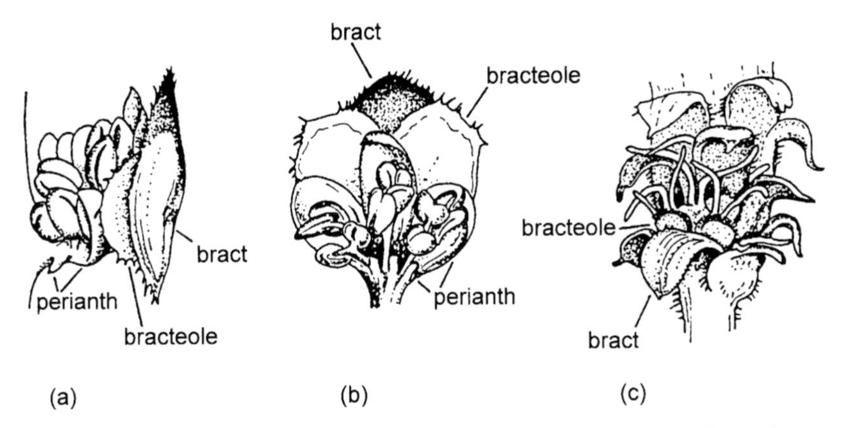

Figure 5.10. Birch flowers (*Betula pendula*). (a) Side view of a group of male flowers borne on a single catkin scale; (b) a similar group seen from below; and (c) part of a female catkin. From: Proctor, M.C.F. and Yeo, P. (1973). *The Pollination of Flowers*. Collins, London. With permission of HarperCollins Publishers Ltd.

on to form the cup of the acorn (Figure 5.9b,c). Just poking out above these bracts is all that is left of the petals and sepals: a green toothed perianth surrounding the three styles. Other wind-pollinated trees have their flowers in even more pronounced catkins ('little cat' after the resemblance to a kitten's tail). Some put just the male flowers in catkins (including hazel and alders), others use them for both sexes (hornbeam, birches, poplars and walnuts). Even though they look so strange the catkins still contain 'normal' flowers. For example, in birch (Figure 5.10) the male catkin is made up of a series of scales or bracts or smaller bracteoles below which nestle three flowers, each with two deeply divided stamens. The remains of the perianth are there but are very small and insignificant. The female catkin is made up in a very similar way. A catkin is really a branch with leaves (bracts) and flowers that has been condensed down into a small flexible dangling branch: a perfectly adapted way of getting pollen into the wind.

Conifer cones (botanically called strobili) are built in a similar way to catkins. A central stem or axis bear the (usually) spirally arranged scales (really modified branches), and it is these scales that carry the male or female parts. In most conifers the female cones have two ovules on each scale, which will become seeds. In the ephemeral male cones (which drop off once they have shed their pollen), the papery scales bear pollen sacs. The ovules in conifers are 'naked', unprotected by an ovary and so with no stigma and no style.

Sending pollen by air

Wind pollination does, of course, require a lot of pollen. Birch and hazel can produce 5.5 and 4 million grains *per catkin*, respectively (pollen production is affected by many factors such as weather, and some trees like ash and elm follow three-year cycles of maximum pollen production). There are various adaptations to help as much of the pollen go as far as possible. Most deciduous wind-pollinated trees produce their pollen while the canopy is bare of leaves to reduce the surrounding surfaces that 'compete' with the stigmas for pollen. Evergreen conifers have less to gain from spring flowering and indeed some flower in the autumn (true cedars, *Cedrus* spp.) or winter (northern incense cedars, *Calocedrus* spp. and the coastal redwood, *Sequoia sempervirens*). They improve pollen dispersal by placing the cones on the ends of branches.

Pollen produced higher in the canopy is likely to go further: it is windier (and gustier) and the pollen can be blown further before hitting the ground (however, conifers tend to keep male cones lower to help prevent self-pollination; see below). Moreover, in dangling catkins like hazel, the scales touch and so hold the pollen in until the wind is strong enough to bend the catkin, ensuring that pollen is only shed into the air when the wind is blowing hard. Weather is also important. Pollen is shed primarily when the air is dry to prevent too much sticking to wet surfaces or being knocked out of the air by rain. Despite these adaptations, much of the pollen fails to leave the canopy and only between 0.5 and 40% gets more than 100 m away from the parent. But once this far, significant quantities can go a kilometre or more. Indeed, pollen can travel many thousands of kilometres at high altitude although this is undoubtedly of such low density that it has little value in pollination. Since these are all floating around in the air, it is no wonder that wind-pollinated trees are a major source of hay fever.

Once the pollen has been snatched by the wind, the fate of the pollen is obviously up to the vagaries of the wind, but not everything is left to chance. Wind-borne pollen is dry, rounded, smooth and generally smaller than in insect pollinated plants to help it fly through the air (20–30 μm diameter in wind pollinated hardwoods, 50–150 μm in conifers[1] compared with 10–300 μm in insect-pollinated plants). But size is a two-edged sword. Small grains may be blown further but they are also more prone to be whisked round the waiting stigma because smaller particles tend to stay entrained in the streamlines that flow around the stigma. But stigmas create turbulence and this 'snow fence' scenario may help pollen stick in the low air speed of the turbulence. Moreover, pollen grains acquire a strong positive charge as they fly through the air, and female flowers have a negative charge, which may also help pluck pollen from the streamlining air.

Conifers don't have sophisticated stigmas to catch pollen, but most do produce a sticky 'pollination drop' at the tip of each scale. Pollen sticks to the drop (probably aided by turbulence) and is pulled down towards the young unfertilised seed as the drop dries. This may be less passive than it seems because if pollen is added to a pollination drop it disappears within ten minutes while others remain unchanged for a few days to a few weeks. Incidentally, the pollen of

[1] 1 micrometre (μm) is a millionth of a metre or 0.001 mm.

pines and some other conifers have two air-filled bladders or sacs on the sides, which are said to help the flotation of the relatively large grains in the air. But settling rates for pollen are the same for conifers and hardwoods—3–12 cm per second on average—and the sacs are reported to shrink in dry air anyway. It may be that the sacs are more important in orientating the pollen grain on the pollination drop. Although pollination drops are found in most conifers (including primitive conifer-relatives such as cycads, the ginkgo and Gnetales; see Chapter 1) they are not found in true firs (*Abies*), cedars (*Cedrus*), larches (*Larix*), Douglas firs (*Pseudotsuga*), hemlocks (*Tsuga*), the southern kauris (*Agathis*) or monkey puzzle relatives (*Araucaria*). Instead they rely on a variety of shaped cone scales to effectively trap pollen as it drifts past. Douglas fir, for example, has a slippery slope leading right down to the ovule, sending any falling pollen on a helter-skelter ride down into the depths of the cone away from any wind that might try and tear the pollen away.

Blurring between animal and wind pollination

Although the distinction between wind and animal pollination, with their different sorts of flowers, appears to be clear-cut, it is likely that in many trees there is a balance between the two. Pollen from all the best-known European wind-pollinated trees—oak, beech, ash, birch, hazel—has been found on honey bees. And even trees normally thought of as being insect-pollinated and which produce abundant nectar—limes, maples, sweet chestnut and eucalypts—release appreciable quantities of pollen into the air. These cases can be thought of as 'accidental leaking' of pollen to the insects or wind but undoubtedly really represents an adaptation to avoid putting all the tree's eggs in one basket. This is not a static situation: as environmental conditions change so the balance of advantages of wind versus animal pollination can change, such as seen in willow on the cusp of wind, insect and bird pollination. Nor is this blurring restricted to hardwood trees. The primitive conifer-like cycads and welwitschia (*Welwitschia mirabilis*; see Figure 2.1d) have a pollination drop and appear to be wind pollinated but it is possible that the pollination drop acts as nectar to attract insects to the female cones while pollen attracts them to the male cones.

The problem of being large

A large animal-pollinated tree in full flower faces distinct problems because of its size. It needs to produce enough reward in its flowers to attract the pollinators but runs the risk of producing so much within one canopy that the pollinators linger rather than go on to the next tree. Moreover, if all other individuals of a species are in flower (which they would need to be to ensure cross-pollination) there may not be enough pollinators to go around the huge number of flowers produced.

One solution to these problems is to be in flower when other species are not so that you have the pollinators to yourself. In temperate regions this is constrained by having to work around winter, but flowering of insect-pollinated trees is indeed spread over a large part of the summer. But this still does not solve the problem of encouraging pollinators to leave the abundant food of one tree to go to another. Some large tropical trees get around this by having some parts of the tree in flower, others in bud and others in fruit, giving pollinators the impression of a series of small trees. A more extreme answer, used by other tropical trees, is to produce just a few flowers at a time over a long period, possibly all year. Hawkmoths, hummingbirds, bats and, especially, solitary euglossid bees exploit these extended blooming trees by 'traplining'. Here they fly over complex feeding routes repeatedly visiting widely spread flowers, just like a trapper in the Arctic visiting his spread-out line of traps. Bees can travel more than 20 km during these daily excursions (by comparison, temperate bees move pollen comparatively small distances: one study showed that 80% of flights were less than 1 m and 99% less than 5 m). This may not be entirely for food; it has been suggested that the bees show territorial behaviour involving scents taken from plants.

At the other extreme, however, are 'mass blooming' trees producing a huge number of flowers. This is especially seen in the aseasonal (no regular dry period) tropical evergreen forests of Borneo and Malaysia where the tallest canopy trees (mostly species of dipterocarp) may all flower and fruit synchronously at intervals of 2–10 years. 'Over a period lasting a few weeks to a few months, nearly all dipterocarps and up to 88% of all canopy species can flower after years of little or no reproductive activity … . The region over which such a mass-flowering event occurs can be as small as a single river valley or as large as northeastern Borneo or peninsular Malaysia.' (Ashton *et al.* 1988). With individual dipterocarp trees presenting up to 4 million flowers, this mass flowering seems to ignore the above problems. But as is often the case, the solutions are subtle and not immediately obvious. Firstly, the dipterocarps flower sequentially, in the same order each time (Figure 5.11), thereby reducing competition for pollinators. Secondly, although an individual tree can flower over 2–3 weeks, individual flowers may last only a day. The main pollinators are small thrips attracted by the overpowering scent of the opening flower to spend the night tramping around inside a flower eating it and the pollen. In the morning the flower falls, complete with its happy thrips (leaving the ovary on the tree to develop into a fruit). As the next wave of flowers opens the following evening the thrips fly up for their next feed, blown by even light winds to, hopefully, land in the flower of a different tree to the night before. Once there they deliver their load of pollen as they wander around feeding. This works only because, unlike most tropical trees, the dipterocarps tend to grow clumped together (the heavy winged seeds of dipterocarps spin like maple 'helicopters' but do not go far). Where do all the necessary thrips come from? The thrips persist at low levels between mass flowerings and explode in numbers as the trees come into flower. Lastly, cross-pollination

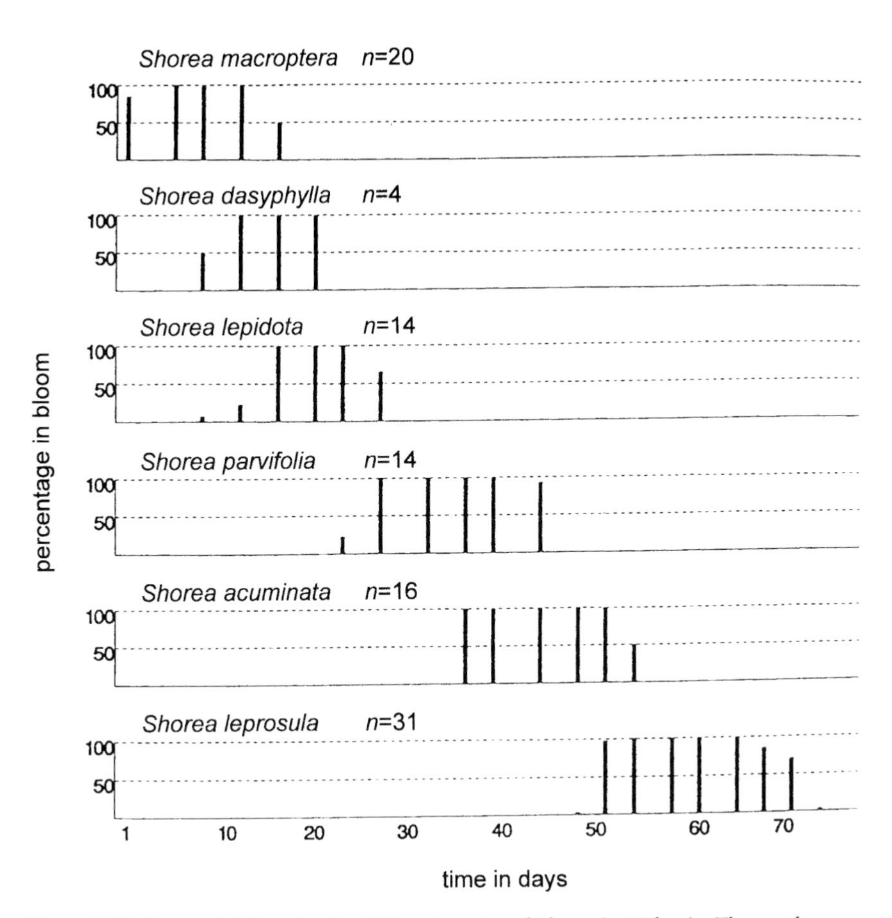

Figure 5.11. Sequential flowering by different species of *Shorea* in Malaysia. The graph shows percentage of trees in bloom on a given day after 14 March 1976 for six species (*n* is the number of trees looked at in each species). From: LaFrankie, J.V. and Chan, H.T. (1991). Confirmation of sequential flowering in *Shorea* (Dipterocarpaceae). *Biotropica*, 23, 200–3.

is encouraged by many of the dipterocarps being self-sterile or even having separate male and female trees (see below).

In general, mass blooming trees use relatively unspecialised pollinators, which gives them a greater number of pollinators to go round. Their strong visual image, created by mass flowering, can attract pollinators in large numbers from over great distances. Up to 70 species of bee have been observed visiting a single tropical tree; such high rivalry can work to the benefit of the tree because in the ensuing battle for flowers some bees can end up being ejected and going to the next tree! But there may be even more subtle ways of encouraging insects to move from one tree to the next. It has been suggested that trees are a continually changing source of quantity and quality of nectar and bees regularly sample nectar from different trees to find the best current source. Why go to this trouble? The answer seems to be that mass blooming and fruiting can be an important mechanism for reducing the predation of flowers and seeds (see page 147).

Wind-pollinated trees do not entirely escape the problems associated with mass flowering. Wind dispersal of pollen is not affected by the number of flowers but the big danger is swamping a tree's flowers with its own pollen. Even if self-sterile, the stigmas could get so coated with the tree's own pollen that other pollen would physically not be able to reach the stigma. In those trees where self-pollination is possible, too much pollen may still be counterproductive. It has been experimentally shown in walnuts that adding extra pollen to flowers resulted in fewer fruits and seeds: too much of a good thing!

Self- and cross-pollination

It has been assumed so far that cross-pollination is better than self-pollination. This perhaps needs to be justified. In the long term it is better for trees and other plants to be cross-pollinated (i.e. pollinated with pollen from a different individual) to prevent inbreeding and the expression of harmful genes. Studies have demonstrated the many effects of 'inbreeding depression' such as poor germination, reduced survival of seedlings, chlorosis (lack of chlorophyll) of seedlings and reduced height growth. Indeed 'hybrid vigour', where a hybrid is more vigorous than either of its parents, may partly be an expression of this outcrossing in comparison to the parents, which may be inbred to an unknown degree (see Sedgley and Griffin 1989). Having said this, self-pollination is a useful short-term way of producing some seeds if no other pollen is available. The common solution is to aim for cross-pollination but fall back on self-pollination if all else fails. The mechanics of doing this are varied, as discussed below.

Self-incompatibility

Some trees are completely 'self-incompatible' and so cannot pollinate themselves. This is particularly common in wind-pollinated trees where lots of 'self' pollen is likely to be blowing around. Thus in beech self-pollination only leads to empty nuts, and well-spaced beech trees produce fewer nuts than those in woodlands. But self-incompatibility is also found in insect-pollinated trees such as apples. Most apples are cloned by grafting and so an orchard of one variety is, in effect, all the same tree; the solution is to plant two or more other varieties. This explains why a lone apple tree may be barren (unless it has several varieties grafted onto one stem, while a single Victoria plum (self-fertile) fruits well. The majority of trees, however, have at least some capacity for self-pollination as an insurance policy. The European mountain ash (*Sorbus aucuparia*), which as the name suggests can live in the hostile uplands, has widely spread stamens in warm weather, exposing the abundant nectar to attract pollinators but, in dull weather, when few insects are around, the stamens converge, leading to self-pollination. Others, by a variety of physical and chemical means, favour pollen from another tree but will often accept their own pollen if no other is available.

Producing the sexes at different times

A common mechanism in trees to reduce or prevent self-pollination is to mature the male and female parts of the flower at different times (called dichogamy). Predominant in wind-pollinated plants (but also found in, for example, the magnolias, pollinated by insects) are flowers where the female stigma is receptive to pollen before the stamens start producing pollen (protogyny). The advantage is obvious: the stigma can catch 'foreign' pollen before being besieged by its own abundant pollen, but can be self-pollinated if foreign pollen is absent. The opposite situation, where the male stamens produce their pollen and wither before the stigma becomes receptive (protandry), is more common in animal-pollinated flowers. Self-pollination here is more tricky but can be achieved by such mechanisms as the bending or curling of the stigma to touch parts of the flower still dusted in its own pollen, or having a late-maturing set of anthers to do the job.

Male and female parts in separate flowers on the same tree: monoecy

Where the separation of the sexes in time becomes so great that either the male or female parts occur too late in the flower's life to work very well, it leads to a functional separation of the sexes into different flowers. Thus we move from having 'perfect' flowers (hermaphrodite flowers with the two sexes in one flower, as in limes, elms, and horse chestnuts) to 'imperfect' flowers holding just one sex. Such separation (called dicliny) is particularly common in wind-pollinated trees and in some with relatively unspecialised insect pollination. For most animal-pollinated trees such separation is too expensive since pollinators will only be doing half the fertilisation process in any one flower—collecting or delivering pollen—effectively doubling the costs of attracting animals.

Although the male and female flowers are separate, they can be very close. In chestnut (*Castanea*) species, for example, the sexes are normally in separate catkins but female flowers can occur at the base of an otherwise male catkin. Normally, however, there are male and female flower heads, which may be on the same branch or in different parts of the tree. This is taken to an extreme in conifers, where the female cones tend to be high up in the crown and the males low down (though it is not always so clear cut). Self-pollination is reduced since pollen seldom moves directly upwards but turbulence carries pollen upwards into the canopy of surrounding trees, helping cross-pollination. And the heavy seeds, being high up in the canopy, are likely to spread further in the wind.

Male and female flowers are not always produced in different parts of the canopy purely for reasons of effective pollination and seed dispersal. In tropical trees especially, flowers and fruits can spring directly from large leafless branches or the trunk (a condition called cauliflory). This is common in trees with large fruits that need a solid support, such as the cocoa tree and the durian. The Judas tree (*Cercis siliquastrum*) is one of the few cauliflorous species outside the tropical forest; here the inflorescence springs from an old leaf scar or from very dwarf shoots just on the bark. Just why is a mystery.

Unisexual flowers can be either on the same tree (monoecious: both sexes in 'one home') as described above, or different trees (dioecious: the sexes in 'two homes', i.e. two trees).

Why be dioecious?

A moment's reflection shows that dioecious trees (having separate male and female trees) are apparently at a disadvantage. True, it ensures cross-pollinating but only roughly half the trees (the ones with female flowers) can produce seeds. Moreover, if the individuals are widely spaced the chances of breeding are much reduced: for example, there is no seed production in relict populations of juniper (*Juniperus communis*) in Wales where the remaining males and females are widely separated. Poor seed set by the dwarf shrub cloudberry (*Rubus chamaemorus*) at the southern limit of its growth in the Pennines of England is considered to be due to the presence of large single-sex clones and a greater frequency of male plants. Dioecy can also make pollination more difficult. Since only half the trees produce pollen, nectar needs to be the main reward, cutting down the potential number of pollinators (and, as noted above under monoecy, animal pollination is more expensive with unisex flowers). Female willow catkins contain three times as much nectar as males and the bird pollinators, blue tits, visit mainly female catkins; presumably the odd visit to a male catkin and most visits to female plants is the most efficient use of the pollinator.

Despite these apparent disadvantages dioecy is particularly common among trees. Only about 4% of all flowering plants (angiosperms) in the British flora, and the world, are dioecious but in the temperate forests of N Carolina dioecy is found in 12% of trees (26% if shrubs are included), rising to about 20% (32% with shrubs) in Costa Rica, and 40% for Nigeria. Dioecy is scattered through many families but only a few are entirely dioecious, notably the Salicaceae: willows, aspens and poplars. Dioecy is found in the ginkgo, cycads, yews, many junipers, the kauri pine of New Zealand, the yellow-woods (*Podocarpus* spp.), about 15% of maples, the temperate holly (*Ilex aquifolium*), tree of heaven (*Ailanthus altissima*), the shrubby butcher's broom (*Ruscus aculeatus*), and many tropical trees. But dioecy is not found in the pines or any members of the family Taxodiaceae (including the redwoods, the Japanese red cedar and Chinese firs) or the monkey puzzle and cypresses.

So why be dioecious? There are several possible answers. Guaranteed outcrossed offspring ensures the highest genetic quality, and dioecy is the only way to absolutely ensure this. A more convincing answer, favoured by Darwin, comes from the well-established link between dioecy and the production of large fleshy fruits containing large seeds. Such fruits are costly to produce (see below) and it may be that trees that do not produce pollen can invest more heavily in bigger seeds and nutritious fruits, and are likely to be more successful (see Chapter 8). Finally, dioecy may reduce seed predation because if only half of the trees of a species are producing seeds, they will be harder to find. Discussion on how and why dioecy arose is still being hotly debated; see Bawa (1994) and Freeman *et al.* (1997) for detailed arguments.

Whatever the reason, the dioecious habit can be useful to people. Male poplars are planted in urban areas to avoid the troublesome fluffy windblown fruits, and female sumacs (*Rhus typhina*) are planted to ensure showy fruit clusters instead of the dull green male blossoms. Female crack willows (*Salix fragilis*) are usually planted along British rivers because they are supposed to pollard better than males.

Trees that swap sex

When you think that single-sex flowers are formed by the loss of one sex or the other, it is perhaps not surprising that there are trees that slip back into old ways. Monoecious trees, normally with unisexual flowers, can sometimes produce hermaphrodite flowers. Dioecy (having male and female flowers on different trees) is rarely absolute. For example, a male or female tree may have a few hermaphrodite flowers (as has been seen in the shrubby butcher's broom and the N American gambel oak, *Quercus gambelii*), or flowers of the opposite sex: yew (*Taxus baccata*) may occasionally produce the odd flower or whole branch of the opposite sex. Others go further and change sex completely, once or repeatedly. Male trees of the normally female Irish yew (a clone originally from one female tree: *Taxus baccata* 'Fastigiata') and female trees of the normally male Italian poplar clones (*Populus* × *euramericana* 'Serotina') are known. Cycads in cultivation have also been reliably reported to change sex, usually following a period of stress such as transplanting, damage, drought or frost.

Others are yet more complicated and seem to write the rules as they go along. The persimmons (*Diospyros* spp.) from N America and Asia can have a few branches of the opposite sex on an otherwise unisex tree; some are consistently monoecious (producing flowers of both sexes) while others produce both sexes some years but not others; and (more rarely) a few hermaphrodite flowers may be produced on otherwise male or female trees. The European ash and some maples take some beating in their apparent total sexual confusion. In the European ash 'some trees [are] all male, some all female, some male with one or more female branches, some *vice versa*, some branches male one year, female the next, some with perfect flowers' (Mitchell 1974) and the variations may differ on a single tree from year to year.

The cost of sex

All this sexual swapping around may seem strange to those reared on human genetics where sex is determined by X and Y chromosomes and cannot be readily changed in an individual. In trees and other plants sex is not as genetically fixed as in higher animals and is frequently affected by environmental conditions (perceived perhaps through the quantity of food stored in the tree). Indeed the proportion of female flowers on monoecious trees generally increases with better growing conditions and increasing age (younger and smaller plants tend to be mostly or wholly male). This reflects the larger cost involved in maternal reproduction; larger, older and better-growing trees have more resources available to invest in the expensive production of fruits and seeds. Changing from male to female is also a way of going out with a bang. The snake-bark maple (*Acer pensylvanicum*) growing around the Great Lakes of N America invades gaps in forests. As the young trees are gradually shaded by other more vigourous trees they change from male to female, putting all their remaining resources into seeds; it is now or never!

The cost of femaleness is also seen in dioecious species. Trees such as yew, ginkgo and poplars show 'male vigour': the males tend to be taller, flower at an earlier age and live longer because of lower reproductive effort.

In the perverse way of nature, there are examples of trees that are predominantly female early on in life. This is true of the New Caledonian pine (*Araucaria columnaris*), a relative of the monkey puzzle, a few pines and young individuals of the N American bigtooth maple (*Acer grandidentatum*). These trees are wind-pollinated and since removal of pollen needs high winds and deposition lower speeds, perhaps short young trees tend to be better at accumulating pollen and do better as females.

Seeds without pollen: apomixis but not parthenocarpy

It is possible for seeds to form without pollen (apomixis). This results in the offspring being identical to the mother. The rose family (Rosaceae) is rich in trees predisposed to apomixis such as species of *Amelanchier* (mespils), *Crataegus* (hawthorns), *Sorbus* (whitebeams, etc.) and *Malus* (apples). Strictly speaking all but the last need pollen for the seed to develop properly but all the genetic material comes from the mother. Regardless of the mechanics, the result is a stand of trees with all individuals looking the same, but most likely slightly different from the next group down the valley. The result is endless discussion about which are separate species and which are merely large clones originating from one mother.

Parthenocarpy is similar (there is no fertilisation from pollen, and the fruit still grows, as in apomixis) but here no seeds develop ('parthenos' meaning virgin). Many common temperate trees (including maples, birches, ashes, elms, hollies, firs and junipers, but not beeches or oaks) are known to produce fruits without seeds. Parthenocarpic fruit are usually smaller than normal but this has been exploited to produce seedless varieties of clementine, navel oranges, bananas, apples and pears. Species that are not parthenocarpic can be encouraged to be so if they are given hormones to replace those normally produced by the seed; hence we have seedless varieties of such things as cherries, figs and mangos (see Sedgley and Griffin 1989).

From flower to fruit

In most hardwood trees, the ovules (young seeds) are fertilised within a matter of hours or days of the arrival of the pollen on the stigma. During this time the pollen grows a threadlike tube through the style and down to the ovule, a journey of a matter of centimetres. In conifers the pollen does the same in the more open environs between the cone scales. Incidentally, pollen grains of cycads and the maidenhair tree (*Ginkgo biloba*) release motile sperm cells, which must swim to the ovule. This is a evolutionary remnant typically associated with more primitive plants such as ferns and horsetails.

The time taken from the formation of flower buds (normally the autumn before they open; see Chapter 6) to the release of ripe seeds varies tremendously. Elms are very quick, shedding their seeds in May to July, just 8–10 weeks after flowering in early spring. The majority of temperate trees, however, disperse their seeds in the autumn. Thus, the whole cycle from formation of the flower buds to release of seed takes around one year. Conifers such as western red cedar (*Thuja plicata*), true firs, larches, spruces, incense-cedar (*Libocedrus decurrens*) and the coastal redwood produce seeds in the same time span but the delay between pollination and fertilisation is often three weeks to two and a half months, which means that, in some of the spruces, seeds develop fully in only 2–3 months.

Other conifers, including most pines, the true cedars, false cypresses (*Chamaecyparis* spp.), junipers, and the giant sequoia (*Sequoiadendron giganteum*) as well as a few hardwoods (notably a number of N American oaks, including the scarlet and red oaks, *Quercus coccinea* and *Q. rubra*), take two years from flower bud to seed fall (Figure 5.12). In pines, fertilisation takes place usually in the spring, one year after pollination. A number of pines delay seed fall even longer but this is usually linked with 'serotiny' where seeds are held in the cone for sometimes decades until released by the heat of a fire (see Chapter 9).

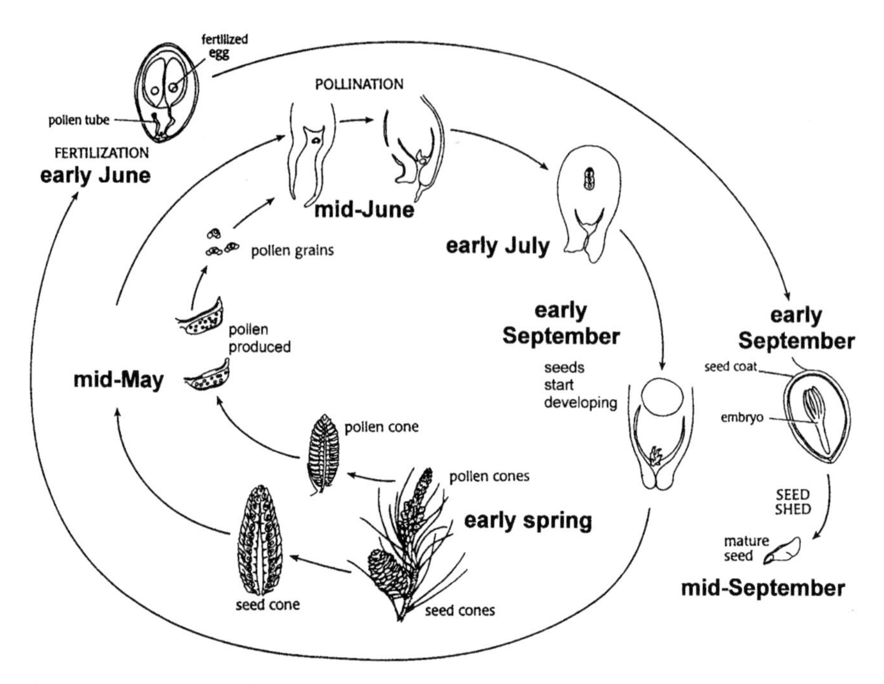

Figure 5.12. The timetable of reproduction in a pine, starting in early spring one year and ending almost two years later. Modified from: Ledig, F.T. (1998). Genetic variation in *Pinus*. In: D.M. Richardson. *Ecology and Biogeography of* Pinus. Cambridge University Press.

Types of tree fruit

As mentioned in Chapter 1, the most fundamental difference between conifers (really the gymnosperms, which literally means 'naked seeds') and hardwoods (really the angiosperms) is the type of fruit.

The scales of conifer cones can be bent apart to reveal the seeds without physically breaking anything apart: the seeds are naked. This is true even in the fleshy or fibrous cones of the junipers. The yew berry is really a bare seed with a fleshy red outgrowth from the base of the seed (an aril; see later) and the fleshy ginkgo 'fruit' has made the seed coat fleshy, so it is still a naked seed.

In hardwoods the seed is completely enclosed by the fruit and cannot be seen without breaking into the fruit, whether this is a dry nut or a fleshy plum. This is because after fertilisation the ovules become the seeds (Figure 5.1) and the ovary becomes the fruit, completely surrounding the seeds. Figure 5.13 shows how hardwood tree fruits are classified and the characteristics of each. You will see that everyday language tends to blur the botanical distinctions between the different fruits. For example, the 'stone' in a date is really just the seed, whereas the stone of a plum is the seed plus a hard inner part of the fruit (which can be cracked open to reveal the seed or kernel). Similarly, many things are sold as 'nuts': some are, such as hazel nuts, but walnuts and almonds are, strictly speaking, drupe stones where the fleshy part of the fruit has been removed (it's still there in pickled walnuts) leaving the hard inner part of the fruit containing the seed. Brazil 'nuts', on the other hand, are really seeds with a very hard seed-coat which grow by the dozen inside a hard woody fruit, which is usually thought of as a berry! Another source of confusion is the alder, which appears to produce cones like a conifer. The cone of an alder is really a woody catkin, and when the 'seeds' fall out they are really complete fruits (small nuts or nutlets) with a dry fruit completely enclosing the seeds.

What does the fruit do?

The fruit has two roles: to protect the seed while it develops and sometimes to help disperse the ripe seeds.

Seeds removed from a tree before they are ripe and able to germinate are a waste for the tree. Moreover, there are many insects angling to eat the nutritious seeds as they develop. In consequence, many fruits (and seeds) are poisonous or astringent when under-ripe, or lacking in nutrients and sugars and so are unattractive to eat (hence the sourness of unripe apples and the high tannin content of unripe persimmons). As the seeds mature and the fruit ripens, the flesh softens, storage materials such as starches and oils convert to sugars, astringent compounds decrease, and to advertise these changes the skin colour changes[2]. Some fruits, however, will retain some of their poisonous or unpleasant characteristics, which reduces the chance of being eaten by the wrong animal.

When it comes to seed dispersal, you will see in Figure 5.13 that succulent or fleshy fruits have a definite role in aiding seed dispersal by animals. Dry fruits can also aid dispersal by forming, for example, the wing of a maple 'helicopter' (samara). Others do little except add an extra dry layer over the seed. Still others, especially those dry fruits containing more than one seed, play no part; they stay on the tree and release the seed to make its own way (technically the fruits that shed seeds are called dehiscent and those that keep the seed inside are indehiscent). Thus the pea-like pods of laburnum split open on the tree, showering the seeds to the ground.

Sometimes where the fruit is not very appetising or nutritious, the seed itself grows a colourful protein-rich covering called an aril. In the yew this forms the bright red 'berry' around the naked seed. In hardwoods the brightly coloured aril can be seen when the fruit opens, as in the mace around a nutmeg and the

[2] Colour in a fruit may be throughout, just in the juice (as in a blood orange) or just in the skin (apples). Two pigment groups give most colours, anthocyanins (red, purple, blue) and anthoxanthins including carotenes (pale ivory to deep yellow). Pigment concentrations are often responsible for different colours: 'black' cherries are actually purple and 'white' actually pale red, while yellow cherry cultivars lack the red anthocyanin in the flesh and have a yellow carotene in the skin.

(a)

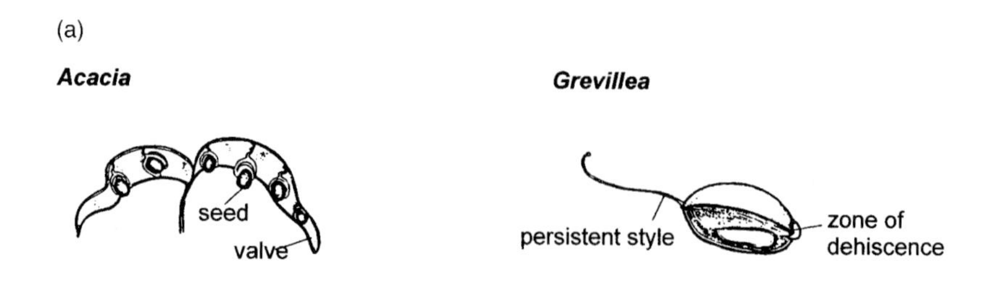

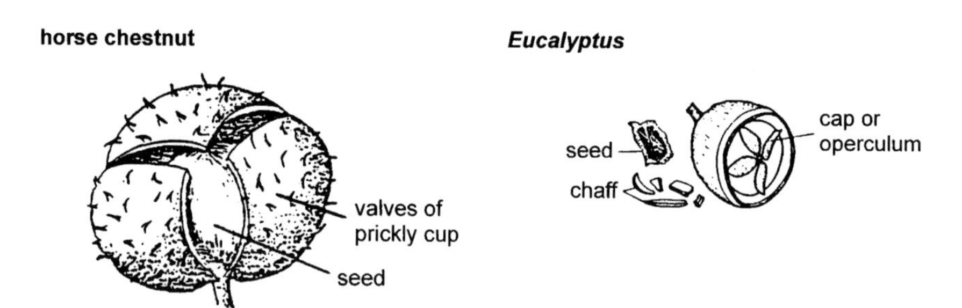

Figure 5.13. Fruit types found in hardwood trees. Note that the fruit (botanically the pericarp) is made up of three layers. In dry fruits they appear as one but in fleshy fruits the outer skin (epicarp) encloses the fleshy layer (mesocarp), and the inner layer (endocarp) can take a variety of forms being hard and woody in drupes to juicy in a berry. From: Sedgley, M. and Griffin, A.R. (1988). *Sexual Reproduction of Tree Crops.* Academic Press, London; Brown, W.H. (1935). *The Plant Kingdom: A Textbook of General Botany.* Ginn & Co., Boston; and Jepson, M. (1942). *Biological Drawings with Notes. Part 1.* John Murray, London.
(a) **Dehiscent dry fruits.**
Legume. The usual fruit of the pea family; it splits into two halves revealing the seeds one on top of the other, e.g. laburnum, gorse, **acacia**. Some tropical species (e.g. sea-bean) break crossways between seeds into small segments.
Follicle. Like a legume but splitting along just one side, e.g. **Grevillea**.
Capsule. Fruit splits along weak lines to let out the seeds, which are side-by-side in the fruit, e.g. willow, horse chestnut, spindle, **eucalypts**, kapok tree.

orange aril in some *Euonymus* species (such as the European spindle tree, where the aril aids dispersal by birds).

Conifer cones are a special case. Here specialised mechanical tissue at the base of each scale controls closing when humidity is high and opening when it is dry. The cones open wider each time, so seed not lost during the first seed fall is released later: 70–90% of seed usually falls in the first two months, the rest falling over winter. To aid the shedding of seeds, most cones gradually bend over from the upright position of pollination to hanging downwards at maturity. Those of

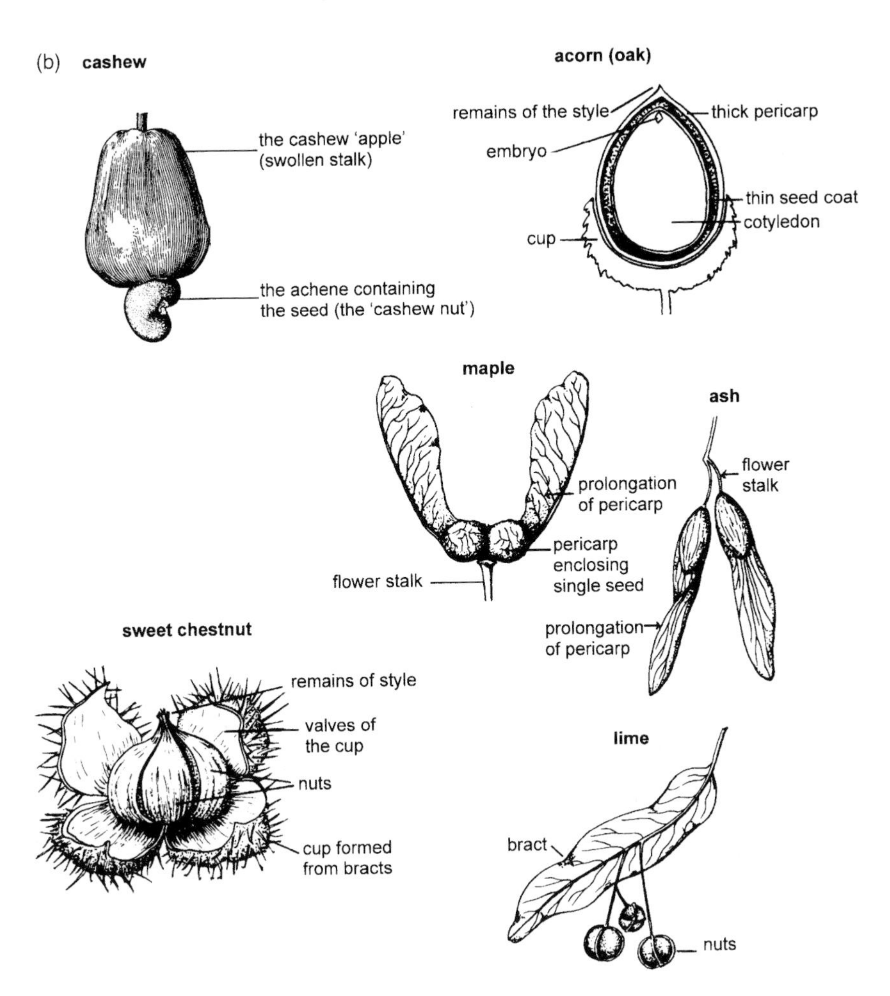

Fig. 5.13 (*cont.*)
(b) Indehiscent dry fruits.
Achene. Single seed with a dry fruit but not as hard and woody as a nut, e.g. **cashew.**
Nut. A common fruit of trees composed of a single seed and a dry fruit that may be hard as in hazel or comparatively soft and easily peeled off as in the **acorn.** A nut may have a bract or bracts (modified leaves) only loosely attached as a wing (e.g. **lime** and hornbeam), or more highly modified and enclosing as hazel, the cup of an acorn, and the hard spiny covering of a beech and **sweet chestnut** (note that these spiny cases do not completely enclose the nut since the stigma and style of the ovary poke out of the end; this separates a nut from the superficially similar capsule of a horse chestnut, where the outer case is the fruit).
Samara. The fruit is elongated into a dry papery wing, as in **ash**, elm, tulip tree and **maple** (note the double wing of a maple fruit is really two samaras joined together).

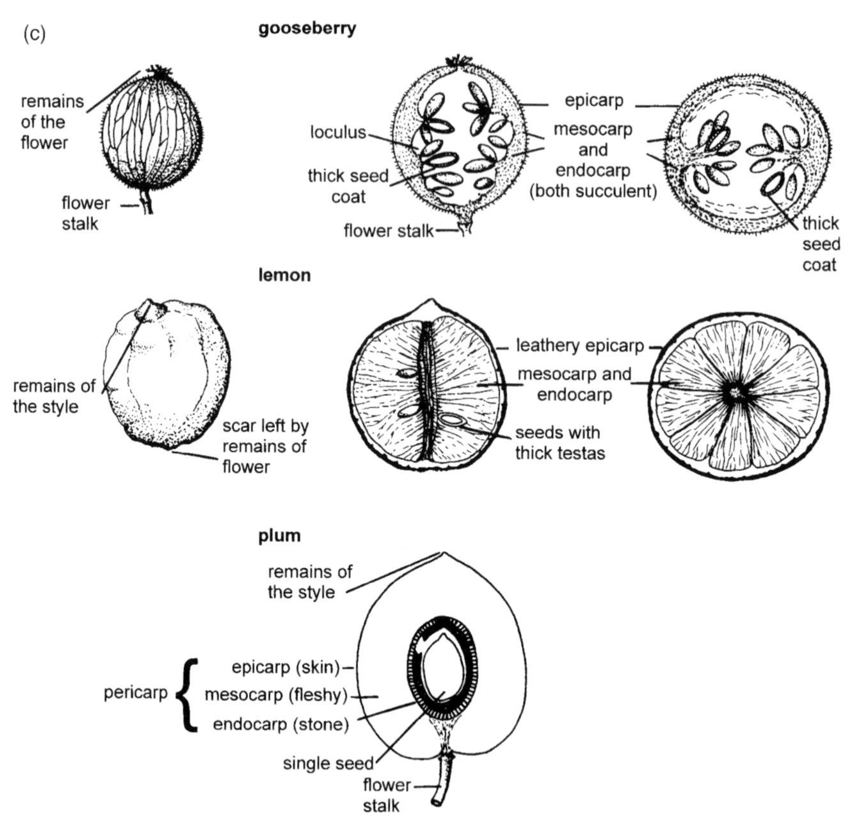

Fig. 5.13 (*cont.*)
(c) **Succulent fruit.**
Berry. A fleshy fruit enclosing the seeds, e.g. **gooseberry**, blackcurrant, orange, banana, bilberry, date (the 'stone' is the seed). In a citrus fruit (called a hesperidium), such as a **lemon**, the outer two layers form the peel and pith, and the inner layer forms fluid-filled hairs or juice-sacs.
Drupe. A fleshy fruit with the innermost layer forming a hard 'stone' around the seed, e.g. **plum**, cherry, peach, almond, walnut, elder, olive. Sometimes we eat the fleshy part and throw away the stone containing the seed, e.g. plums and peaches, or we throw away the fleshy part, crack open the stone and eat the seed, e.g. walnuts and almonds. Some drupes contain more than one stone, e.g. hawthorn, rowan, holly, medlar. The coconut we buy is just the stone (containing solid and liquid food for the germinating seed), and the coir, removed to make coconut matting and peat-free compost, etc., is the equivalent of the flesh and skin of a plum.

the true firs, true cedars and the monkey puzzle relatives are exceptions and stay pointing up. Since their scales are pointing upwards, the only way they can effectively lose seed is for the whole cone to disintegrate on the tree, leaving the central axis on the branch pointing up like a candle. This is why Victorian gentlemen would spend happy hours shooting cones out of the tops of these trees—it was the easiest way to get an intact specimen!

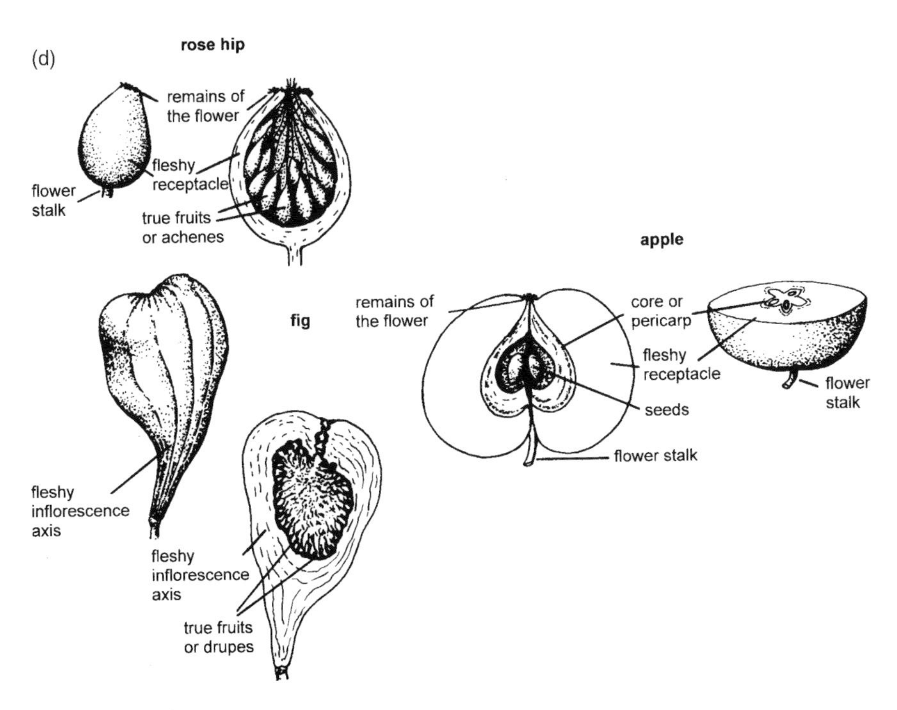

Fig. 5.13 (*cont.*)
(d) False fruits.
Some fruits are supplemented with other structures. In **apples** and pears (pomes) the base of the flower (the 'receptacle' where all the bits like stamens and petals are joined on) grows up and around the true fruit: the core is the real fruit and the bit we eat is the 'fleshy receptacle'. This is similar in **rose hips** (containing achenes) and **figs** (containing drupes) where the receptacle from one (rose) or more flowers (fig) grows up and encases a mass of individual small fruits.
Several flowers can grow together as the fruits develop to produce what appears to be one fruit, e.g. plane, osage orange, mulberry, and the cones of alder and banksia: these are multiple fruits. Strictly speaking the false fruits of fig and the double samara of maples could be called multiple fruits.

Mast years

Tree such as birches, aspens, willows and elms produce a large and fairly constant number of seeds each year (around 250 000, for example, in alder and 20 million in the foxglove tree, *Paulownia tomentosa*). Others, especially those with relatively large seeds (such as oaks, beech and ash) and conifers, show great variation from year to year, often rhythmically. Acorn production in British oak can vary from almost none to over 50 000 and occasionally 90 000 per year. The good years, in which flower and seed production are exceptionally high, are referred to as mast years. Mast years for beech (Figure 5.14) and oak in Britain may be every 2–3 and 3–4 years, respectively, with exceptional years occurring

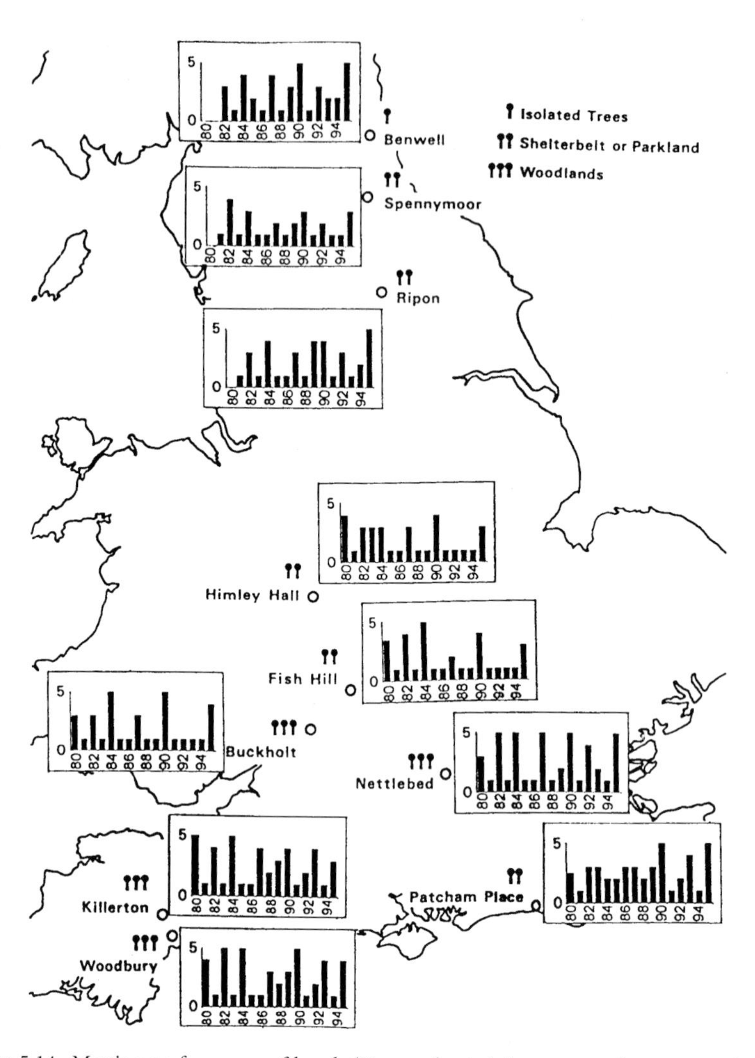

Figure 5.14. Masting performance of beech (*Fagus sylvatica*) for groups of trees around Britain for the years 1980–1995. The amount of mast is expressed on a five-point scale using the average number of nuts collected in a seven-minute period (1, fewer than 10 nuts collected; 2, 10–50; 3, 51–100; 4, 101–150; 5, more than 150). From: Hilton, G.M. and Packham, J.R. (1997). A sixteen-year record of regional and temporal variation in the fruiting of beech (*Fagus sylvatica* L.) in England (1980–1995). *Forestry*, **70**, 7–16. Reprinted by permission of Oxford University Press.

every 5–12 and 6–7 years, respectively. The N American aspen (*Populus tremuloides*) has been recorded as producing a maximum of 1.6 million seeds every 4–5 years. Even apples tend to produce biennially good crops.

Masting can only happen when there are sufficient food reserves in the tree. Large-seeded trees invest an enormous amount of energy into growing seeds and in good mast years a tree puts everything into producing a large quantity of seed, so much so that growth of the tree may be reduced that year. With such effort it is perhaps not surprising that there are rarely two mast years in a row. The tree must rebuild its stored food reserves ready for the next big year, and in those intermediate years it may produce just a modest amount of seed or none. Many factors affect how much food can be accumulated in a tree, including soil, aspect and amount of shading (acorn yields are higher from trees with spreading crowns and conifer seed-yields are increased by thinning). But most variable from year to year is the weather. Growing conditions affect not just how much food can be spared for storage in any one year but also when these stored reserves can be used. Bad weather, either preventing buds from forming or killing them later, will prevent masting even when the tree has ample reserves.

Since weather is the main governing factor (though perhaps not always the only factor), mast years should be synchronous across large areas (Figure 5.14). Generally this is true although local variations in weather will inevitably play a large role in how well even geographically close trees are synchronised, and it is not uncommon for even different parts of southern Britain to be out of phase (e.g. 1987 in Figure 5.14). A late spring frost or attack of herbivores in one area, for example, may prevent trees masting with others around it.

This explains how mast years arise but not why. Why doesn't an oak produce a smaller, more constant and regular number of acorns every year, more like birches and aspens? The answer appears to lie in the fact that the large seeds of masting species such as oak are eagerly sought after by seed-eating herbivores like squirrels and wood pigeons (the latter can take 100 acorns per day). Masting swamps the seed eaters with more seeds than they can consume (predator satiation) leading, hopefully, to a few surviving to germinate. Studies of beech in Britain have shown that up to 100% of beech seed is eaten by mice and birds in years when there is a poor crop but over 50% of seed may be left at the end of a winter following a mast year. This number is left despite the large flocks of bramblings and great tits attracted from mainland Europe in good beech mast years. Oak and beech often only produce seedlings following mast years.

Pinyon pines (*Pinus edulis*) and a number of other pines around the world have evolved large seeds without wings. They are spread primarily by corvid birds in much the same way that jays and squirrels move acorns in Britain (see 'Animal dispersal' below). The birds carry seeds away and bury them to eat during the winter, but in mast years some will be forgotten and left to germinate. In the pinyon pine (at least) the birds do not feed on trees with low seed numbers, so masting acts to *attract* these seed-eating birds.

Janzen (1971) suggests how masting could have evolved. If the weather either destroys a flower crop or fails to provide a flowering cue one year this leads to a larger crop next year with more of the seed surviving predators and going on to form the next generation. This leads to selection of those plants most sensitive to the disruptive weather event, leading to future masting.

So why don't all trees have mast years? Masting occurs in those trees with largish seeds that are wind-dispersed or where dispersal is via mammals or birds (like beech and oak) that carry off the seed to eat but store some in caches in the ground, which are subsequently forgotten. Conversely, masting is not prominent in trees that spread their seeds by using animal-eaten fruits: it would be self-defeating if fruits were left uneaten. Nor is it common in trees with small seeds such as birch and willow, which presumably do not attract such herbivory. The tropics have an extra problem: in areas where the animals are many and diverse, predator satiation by mast seeding does not occur, presumably because the predators can move and change diets in enough numbers to soak up the increased quantities of seed. But in the animal-poor dipterocarp forests of SE Asia there *is* masting. Although the flowering of different species may be staggered (Figure 5.11), the seed fall is remarkably coordinated to drop large numbers of seeds at the same time. Even with an influx of migrant animals not all the seeds are eaten.

From this discussion it can be seen that a good berry crop is unlikely to be a sign of Mother Nature providing for a bad winter ahead; fruit production is a facet of how good it has been not how good it is going to be, and generally it is the summer before the current one that is responsible for how good the flower crop is.

Dispersing the seeds

On the whole, trees use the same methods of moving seeds as they do pollen. Pioneer trees (such as birches, ash and pines), which obviously tend to grow in open and relatively windy places, have come to rely on wind to disperse fruits and seeds. Trees of woodland are faced with relatively calm conditions and a need for large, heavy seeds (see Chapter 8); consequently they tend to be dependent on the specialised collecting habits of mammals and birds. For similar reasons many woodland shrubs bear fleshy fruits. But there are always exceptions! Seed dispersal in animal-rich tropical forests is primarily by animals, and even among trees that inhabit the windy areas at the top of the canopy, only about half the species are dispersed by wind.

Wind dispersal

Wind dispersal is most effective if the seeds are liberated as high up as possible and have ways of slowing their fall to allow the wind to blow them a long way. Some trees, such as eucalypts and rhododendron, rely on fine dust-like seed. Slowing the speed of fall with larger seeds is accomplished by extending the dry fruit into the samara wing of maples, ashes and tulip trees (*Liriodendron* spp.). If the fruit itself is not much help in flying, a bract can be added as a wing (e.g. hornbeam or lime), or the seed itself can be given a wing (conifers), or a plume of long hairs (willows, poplars, balsa, the Indian bean tree and relatives, and the kapok tree).

Hairs act simply as parachutes but the wings have a more complicated job: they act to spin the seed round, like a helicopter. The size of the wing fits closely to the size of seed. If the seed is too big or small for the size of the wing the whole thing just plummets like a brick (try nibbling bits of a maple wing and throwing it in the air).

Many trees let go of the seed when leafless to allow them to go further: from maples shedding fruits in the autumn to the kapok tree of the New World tropics, which flowers and sheds its plumed seeds when leafless (each tree may only flower every 5–10 years, but can produce 800 000 seeds when it does; it is no wonder that enough kapok could be collected to stuff mattresses and life-jackets before artificial replacements were mass-produced). Trees do not just drop their winged seeds at random; it is much more calculated. As the fruits hang on the tree the separation layer (similar to that developed in leaves in the autumn; see Chapter 2) develops rapidly with low humidity, which, in temperate areas, occurs typically in the early afternoon when wind speeds are highest. The residual amount of force required to remove the fruit ensures that it is plucked off by gusts of high wind, helping it to go twice the distance that one would predict from average wind speeds.

Just how far wind-blown fruits and seeds travel is difficult to calculate (how do you know which tree the seed came from?) and even harder to interpret (how important is the odd seed that travels great distances?). Wind-dispersed seeds are most common in pioneer species like birch, ashes and many pines (although the two North American redwoods are forest species with small winged seeds, which do best on open burnt areas) which you would expect to benefit from spreading evenly over a large area to increase the chance of finding a new bare spot to invade. And this is what happens. Variation in air turbulence is effective at dropping seeds at varying distances from the parent. Most winged fruits and seeds, being relatively heavy, don't go very far, and typically drop within 20–60 m of the parent although there are records of sycamore travelling 4 km and poplars 30 km. Jeffrey pine seeds (*Pinus jeffreyi*) in Nevada have been seen to be carried up to 25 m further once they hit the ground by seed-caching small mammals. Small light seed like birch or heather seeds still in their capsules can be blown many kilometres over hard packed snow to cover an area more than three times that by wind dispersion alone.

Animal dispersal

Although animals can sometimes be inadvertent spreaders of fruits and seeds, as for example when they are caught on fur and feathers and are preened out later, most movement by animals is based on supplying them with food. Seeds are bigger than pollen, therefore insects play a fairly small role: larger animals are needed. Fruits, especially those in the tropics, differ enormously in size, abundance and nutritional properties, consequently utilising a vast range of different animals but especially birds. Fruit attractive to birds tends to be presented in the morning, highly coloured (red fruit in the summer and on evergreen shrubs—wild cherry, holly, yew—and black fruit in autumn—blackberries, elder, bird cherry, ivy—to ensure it stands out) with little smell. Mammals are attracted primarily by smell (e.g. apples, pears). Some fruit is aimed at general feeders and tends to offer a general diet of carbohydrate, which needs supplementing elsewhere. Others, such as the avocado, offer a much more complete package of nutrition and utilise the undivided services of a more restricted number of seed distributors.

Once the animal feeds, the seeds are either spat out or regurgitated (as the splendid quetzal bird of Guatemala does to wild avocado stones), or they are swallowed and have to survive crushing by teeth or gizzards to pass through the gut and appear in the droppings, hopefully some way from where they started. Elder saplings are often seen below starling roosts. The journey through the animal can be eased by speed or by extra protection. 'It is no surprise that syrup of figs eases constipation: [strangler] fig trees planned it that way. From the fig tree's point of view, the effect of fig fruit on the bowels encourages its seed to be dispersed widely in all directions' (Mitchell 1987). Protection can come from hard layers such as with the hard inner shell of a drupe. The passage through the animal can be distinctly beneficial; abrasion of the seed coat of holly by passage through the gizzard of birds improves germination.

Seeds with dry fruits, such as nuts, are eaten for the seed itself. This may seem somewhat self-defeating for the tree since a digested seed can't grow. But the system works by depending upon the clumsiness or forgetfulness of animals. Seed may be dropped as they are being carried away, as happens with small birds feeding on European beech nuts and with parrots feeding on the pods of New World *Parkia* trees. Other animals hoard caches of the seeds for later (especially during mast years), some of which are forgotten, providing the seed with a convenient system of being planted ready to germinate. This is seen with various pines around the world, which have developed large wingless seeds moved by corvid birds; the tough woody cannonball-like fruits of the brazil nut tree, which only the agouti can chew into and who then caches the seeds in the ground; acorns moved by jays and squirrels in the UK. Incidentally, acorns rapidly lose moisture and viability when left on the soil surface, so burying by jays is especially important. The grey squirrel introduced to Britain, on the other hand, frequently bites the end off acorns, which prevents them germinating.

Seed movement is not always straightforward. For example, the Australian quinine bush (*Petalostigma pubescens*) has a three-stage dispersal process. The round fruits (2–2.5 cm diameter) are eaten by emus, which strip the flesh from the drupe stone. Once passed through the gut, there may be over a thousand seeds in a single emu scat, which could lead to intense seedling competition. But all is not lost; the stone of the drupe exposed to the sun explodes 2–3 days later and fires the seeds 1.5–2.5 m. The seeds have a conspicuous fatty appendage (elaiosome) at one end and are carried off by ants into nests where the ants eat the appendage and discard the seed, leaving it underground, where it gets protection from mice and fires, and a favourable germination site. Although spectacular, it is not so different from the broom and gorse of Europe where the pea-like pods explosively scatter the seed as they dry in the sun, and the seeds are similarly carried off by ants, and conveniently buried, for a similar fatty appendage.

How far are seeds taken by animals? Most information stems from studies done in Europe where mammals and birds take acorns, hazelnuts and berries of the order of 10–30 m from the tree, although jays may carry acorns several kilometres. The important thing to remember here is that seeds do not necessarily have to go long distances since the likelihood of a gap in the canopy (by, for example, a tree falling over) is as great near as far. And despite the heavy losses expected by eating this can work very well: in 1968 Mellanby found up to 5000 oak seedlings per hectare in a bare field from acorns carried by birds.

Dispersal by water

Trees that habitually grow beside water have evolved corky fruits that float. The small winged fruits have an oily outer coat and are capable of floating in still water for more than a year; they can be moved by running water and wind drift over still water considerably further than they can be blown through the air. A number of tropical legumes that grow beside rivers drop their pods into the water, where they break into small one-seeded segments and float away. Many wash up close by but some go further, even out to sea where they may float for more than a year and still be viable; indeed many of these species are found world-wide. It is not unusual to find 'sea-beans' (seeds from the pods of *Entada gigas*) washed up on British beaches, looking like fat chocolate coins 3 cm in diameter, having travelled from the Caribbean on the Gulf Stream. Water probably holds the long-distance record: coconuts, wrapped in their fibrous coir float with an impermeable skin, can sail the oceans between continents on journeys of over 3000 kilometres.

Over 100 000 km^2 of the Amazon jungle are flooded each year during the wet season just when many trees and vines ripen their fruits. The fruits of rubber trees (*Hevea brasiliensis*) explode on the tree, hurling the seeds 10–30 m out over the water. The seeds have a shell nearly as hard as that of a brazil nut but fish with molar-like teeth, resembling those of a horse, can crack them open. Many seeds are lost in this way but inevitably some survive floating around for the next 2–4 months. Other seeds encased in fleshy fruits (built-in buoyancy aids) are either swallowed by fish or drop to the bottom as the flesh decays; either way the seed is moved. Folklore has it that electric eels wrap themselves around the base of submerged palms and give them an electric shock—up to 500 V—to dislodge fruit (although it is doubtful whether the palms would be conductive enough for this to work).

The cost of reproduction

Producing flowers and fruits is expensive. Young trees can't afford to do it and older trees pay a cost in doing it. Reproduction slows height and girth growth, and ring widths may be halved in good seed years. In Scots pine, cone production has been estimated to reduce the amount of wood grown by 10–15% during an average year. This is despite a tree doing everything to grow as much food as possible. Trees start at a disadvantage because the flowers take the place of some leaves: male cones in lodgepole pine (*Pinus contorta*) result in 27–50% reduction in the number of needles on that branch (female cones, hanging on branch ends, replace few needles). But leaves immediately adjacent to fruits compensate by working harder, increasing their photosynthetic rate by up to 100%. Nor are flowers and fruits helpless. The green sepals of apple flowers have been measured to contribute 15–33% of the carbohydrate needed during flowering and fruit set. Green fruits also contribute to their own growth. This has been measured at just 2.3% in bur oak (*Quercus macrocarpa*) of eastern N America rising to around 17% in Norway spruce (*Picea abies*), 30% in Scots pine (conifer cones are green when young) and 65% in Norway maple (*Acer platanoides*) with its large green samara wings.

Why grow more flowers than are needed?

It is common for trees to produce mature fruits from only a small proportion of their female flowers. Pines may produce mature fruit from over three quarters of their flowers going down to less than 1 in 20 in mango (*Mangifera indica*) and teak (*Tectona grandis*) and only 1 in 1000 in the kapok (*Ceiba pentandra*) in tropical Africa. Part of this failure can be due to lack of pollination. Normally one or more fertilised seeds are needed to stimulate fruit development via hormone release from the seed, and it is not uncommon for fruits and cones to be dropped if they contain too few seeds (the threshold number can vary annually in any one tree). However, even where pollination fails it is sometimes seen that fruits develop anyway (see parthenocarpy, earlier in the chapter).

On the whole, lack of pollination is a small part of the answer. The overriding reason that flowers fail to produce fruit is because the young fruit abort. This may be due to frost or drought damage, or attack by fungi or animals. But there is a lot of evidence that plants choose to discard excess young fruit. Aborting unwanted fruits early is good husbandry since fruits are costly to produce. But this begs the question: why produce unwanted flowers and fruits in the first place? One possible answer is to do with 'plant architecture'. A flower head may be composed of many flowers (possibly to present a good display to attract pollinators), each capable of producing a fruit but for which there is physically not enough room for all to develop.

Perhaps a more general answer is that, firstly, it is a way of producing as many fruits as possible in a good year. This is a similar process to that seen in birds, which produce an optimistic number of eggs and the smallest chick(s) dies if there is not enough food and survives if there is. In the same way a tree can take advantage of good years when there is plenty of spare food to grow extra fruit. Young fruits are 'competing' with new leaf growth for resources, and so in times of shortage fruits may be 'starved to death' (see Chapter 6). Flowers at the base of an inflorescence have first claim on food and water and therefore tend to be the ones that survive, so as with birds it not a case of all starving, but the weakest go to the wall while the others prosper. Secondly, aborting fruit allows the tree to keep the best fruit. Trees may be able to detect fruits heavily infested with seed-eating grubs and selectively abort them. There is also some evidence that cross-pollinated flowers begin fruit-growth quicker than self-pollinated flowers, in which case, the cross-pollinated flowers will obtain first call on resources and be the ones to survive. In this way the tree will grow the preferred cross-pollinated seeds in preference but will fill up with self-pollinated seeds if there are enough resources available.

In a handful of pines, the opposite problem sometimes occurs where far more flowers are produced and survive than is normal. In these cases of 'aggregated cones' there may be more than a hundred cones in a whorl compared with the normal 1–5. The cones are typically smaller than normal but they may still use up so much food that the branch above the cluster dies.

The next generation

What happens once the seeds have been spread around? See Chapter 8!

Further reading

Ashton, P.S., Givnish, T.J. and Appanah, S. (1988). Staggered flowering in the Dipterocarpaceae: new insights into floral induction and the evolution of mast flowering in the aseasonal tropics. *American Naturalist*, **132**, 44–66.

Bawa, K.S. (1994). Pollinators of tropical dioecious angiosperms: a reassessment? No, not yet! *American Journal of Botany*, **81**, 456–60.

Bazzaz, F.A., Carlson, R.W. and Harper, J.L. (1979). Contribution to reproductive effort by photosynthesis of flowers and fruits. *Nature*, **279**, 554–5.

Chambers, J.C. and MacMahon, J.A. (1994). A day in the life of a seed: movements and fates of seeds and their implications for natural and managed systems. *Annual Review of Ecology and Systematics*, **25**, 263–92.

Dick, J. McP., Leakey, R.R.B. and Jarvis, P.G. (1990). Influence of female cones on the vegetative growth of *Pinus contorta* trees. *Tree Physiology*, **6**, 151–63.

du Toit, J.T. (1992). Winning by a neck: some trees succeed in life by offering giraffes a meal of flowers. *Natural History*, August 1992, 28–33.

Freeman, D.C., Doust, J.L., El-Keblawy, A. and Miglia, K.J. (1997). Sexual specialization and inbreeding avoidance in the evolution of dioecy. *Botanical Review*, **63**, 65–92.

Greene, D.F. and Johnson, E.A. (1994). Estimating the mean annual seed production of trees. *Ecology*, **75**, 642–7.

Hilton, G.M. and Packham, J.R. (1997). A sixteen-year record of regional and temporal variation in the fruiting of beech (*Fagus sylvatica* L.) in England (1980–1995). *Forestry*, **70**, 7–16.

Janzen, D.H. (1971). Seed predation by animals. *Annual Review of Ecology and Systematics*, **2**, 465–92.

Jenni, L. (1987). Mass concentrations of bramblings Fringilla montifringilla in Europe 1990–1993: Their dependence upon beech mast and the effect of snow-cover. *Ornis Scandinavica*, **18**, 84–94.

Kay, Q.O.N. (1985). Nectar from willow catkins as a food source for Blue Tits. *Bird Study*, **32**, 40–5.

Kay, Q.O.N. and Stevens, D.P. (1986). The frequency, distribution and reproduction biology of dioecious species in the native flora of Britain and Ireland. *Botanical Journal of the Linnean Society*, **92**, 39–64.

Lanner, R.M. (1996). *Made for Each Other: a Symbiosis of Birds and Pines.* Oxford University Press, Oxford.

Mellanby, K. (1968). The effects of some mammals and birds on regeneration of oak. *Journal of Applied Ecology*, **5**, 359–66.

Mitchell, A.M. (1974). *A Field Guide to the Trees of Britain and Northern Ireland.* Collins, London.

Mitchell, A.W. (1987). *The Enchanted Canopy.* Fontana, London.

Obeso, J.R. (1997). Costs of reproduction in *Ilex aquifolium*: effects at tree, branch and leaf levels. *Journal of Ecology*, **85**, 159–66.

Perrins, C.M. (1966). The effects of beech crops on Great Tit populations and movements. *British Birds*, **59**, 419–32.

Richards, A.J. (1986). *Plant Breeding Systems.* Allen & Unwin, London.

Seavey, S.R. and Bawa, K.S. (1986). Late-acting self-incompatibility in angiosperms. *Botanical Review*, **52**, 195–219.

Sedgley, M. and Griffin, A.R. (1989). *Sexual Reproduction of Tree Crops.* Academic Press, London.

Silvertown, J.W. (1980). The evolutionary ecology of mast seeding in trees. *Biological Journal of the Linnean Society*, **14**, 235–50.

Stephenson, A.G. (1981). Flower and fruit abortion: proximate causes and ultimate functions. *Annual Review of Ecology and Systematics*, **12**, 253–79.

Vander Wall, S.B. (1994). Removal of wind-dispersed pine seeds by ground-foraging vertebrates. *Oikos*, **69**, 125–32.

Chapter 6: The growing tree

A common question is to ask 'how quickly will my tree grow' or 'how big will my tree eventually get?'. In this chapter we will look at these and related questions, and the reasons behind the answers.

Speed of growth

Height

You have probably seen films where the hero in the Orient is strapped over a bed of growing bamboo as a means of torture and eventual death, speared by the hard growing shoots. The reason this works is the extraordinarily fast growth of over half a metre per day. While tropical vines and lianas can grow almost as fast, trees proper can't equal this rate but can nevertheless be impressively quick, especially when young. A number of tropical species can add 8–9 m to their height in a year. A New World relative of the elm, *Trema micrantha*, has been seen to grow 30 m in 8 years (an average of 3.75 m per year) and a eucalypt (*Eucalyptus deglupta*) in New Guinea reached 10.6 m in just 15 months. *The Guinness Book of Records* quotes the air-speed record for a tree as a specimen of *Albizia falcata* planted in Malaysia, which grew 10.74 m (35 ft 3 in) in 13 months! As you would expect, there is a lot of variation between species. Trees that invade gaps in tropical forests, and need to grow quickly to win the race to the top, grow faster (an average of 1.5–4.0 m in height per year) than later species that can afford to slowly plod upwards through the shade (0.5–1.2 m per year).

Temperate trees obviously grow more slowly. Vigorous young trees in Britain usually grow something between 15 and 50 cm per year (a tortoise-like 1–2 mm per day) although 1.5 m is not impossible, and coppice shoots (with their established roots) can grow 3 m in the first year. As in the tropics, pioneer trees tend to grow faster than those that invade ready-grown forests. There's an old Lancashire saying that a willow will be worth a horse before an oak will be worth a saddle: the pioneer willow grows faster than the oak. Growth in any tree is also affected by age. As trees get taller, the growth rate inevitably slows down to just a few centimetres a year before finally stopping.

Measuring minimum growth in trees is much like the slow cycling record, which became pointless when someone learnt to stay stationary! Where the environment is so inhospitable that new growth is negligible or constantly killed back,

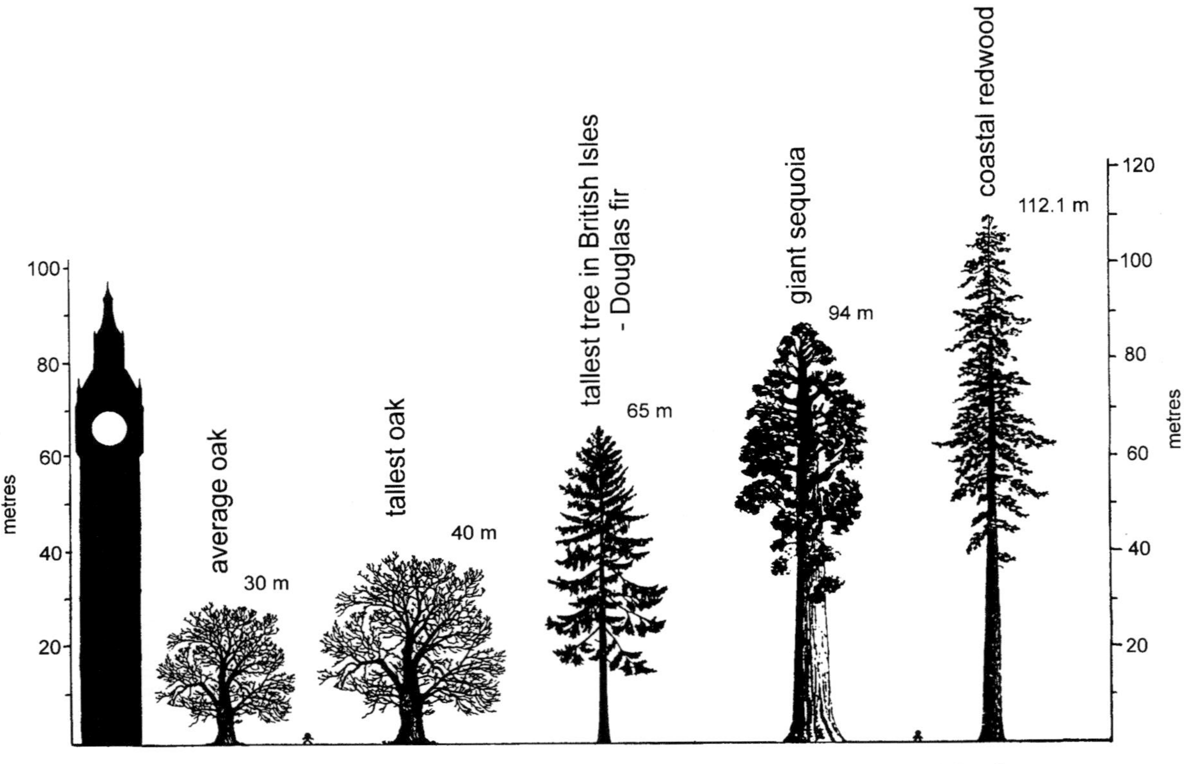

Figure 6.1. Tall trees. An average and the tallest oak in the British Isles give an impression of just how tall the world's tallest trees are compared to Big Ben in London (98 m).

trees may not gain height at all for many years. Trees on northern tree-lines can be hundreds of years old but no more than a few tens of centimetres high.

Width

As we saw in Chapter 3, a tree trunk must normally keep getting wider by forming new rings to ensure water transport up the tree. In temperate trees the average width of new rings is from 0.1 to around 5–6 mm, although trees such as young giant sequoias (see below) can grow such wide rings that tree diameter increases by several centimetres each year. The smallest rings found so far, just 0.05 mm wide, have been measured in three different species spread around the world: white cedar (*Thuja occidentalis*) growing on cliffs of the Niagara escarpment, bristlecone pines (*Pinus aristata*) in California and yew (*Taxus baccata*) in Britain. For these trees it can take 10 years to add 1 mm to the trunk diameter. In the white cedar, an individual little more than 5 cm in diameter was 422 years old! All these examples were growing on very inhospitable dry rocky areas.

Size of the trunk can be useful for ageing a tree, not just be counting the rings, but by measuring the circumference. Alan Mitchell proposed a general and useful rule that temperate trees increase in circumference (girth) at 2.5 cm (1 in) per year when growing in the open, reducing to half an inch in woodland. Remarkably, as a very rough rule, it works (corresponding to an average ring width of 4 mm per year). Generally, young trees grow faster than this, and mature trees more slowly, averaging out to the inch rule except in very young and old trees (more later). There are of course exceptions: a number grow faster, increasing at 5–7.5 cm (2–3 in) per year, including the giant sequoia (*Sequoiadendron giganteum*), coastal redwood (*Sequoia sempervirens*), cedar of Lebanon (*Cedrus libani*), Sitka spruce (*Picea sitchensis*), Douglas fir (*Pseudotsuga menziesii*), southern beeches (*Nothofagus* spp.), turkey oak (*Quercus cerris*), tulip tree (*Liriodendron tulipifera*) and London plane (*Platanus* × *hispanica*). Others grow slower than expected: trees such as Scots pine (*Pinus sylvestris*), horse chestnut (*Aesculus hippocastanum*) and common lime (*Tilia* × *europaea*).

Champion trees in size

Trees are the biggest living things on Earth, no matter how you measure. The tallest living trees are the coastal redwoods (*Sequoia sempervirens*) of the Californian coast: the tallest was measured as 112.2 m (368 ft) in October 1996 (Figure 6.1). It is a remarkably thin tree in proportion to its height, a mere 3.1 m in diameter at the base (although they can be up to 6.7 m). The tallest tree ever is reputed to have been a *Eucalyptus regnans* in Victoria, Australia measured at 132.6 m (435 ft) in 1872 and thought to have been over 150 m (500 ft) at its peak. Uncorroborated records of Douglas firs over 140 m exist. Strictly speaking, the *longest* woody plant is a rattan (a climbing palm) once measured as 171 m

(560 ft) long, but being a vine it has no responsibility for its own support and so doesn't really count.

California also boasts the world's largest living thing in bulk: the giant sequoia (*Sequoiadendron giganteum*) growing in the Sierra Nevada Mountains. The largest individual is General Sherman in Sequoia National Park, which has an impressive height of 83.8 m—although they can be up to 94.8 m (311 ft)—which, combined with a trunk width of 11 m at the base, results in an estimated mass of 2030 tonnes. As an idea of how big the tree is, in 1978 the General dropped a branch 45 m (150 ft) long and nearly 2 m (7 ft) in diameter: a branch larger than any tree east of the Mississippi River. For comparison, blue whales, the largest animals, weigh in at around a mere 100 t apiece! Several fungal colonies found in America and China cover huge areas and may weigh more than General Sherman but although each colony is all genetically identical, are all the bits still connected together as one organism or is it a clone of now separate individuals? In the same way, aspen clones (*Populus tremuloides*)—suckers from the same original tree—can cover vast areas. The biggest reported is in Utah and covers 43 ha (106 acres) with an estimated mass of more than 6000 t, but again, is it still all one tree?

To put these Californian giants into perspective, European and N American hardwood trees commonly reach 30 m and under favourable conditions may reach 45 m or so (see Figure 6.1). The tallest tree in the British Isles as I write is a Douglas fir in Scotland 64.6 m (212 ft) high with a diameter of 1.32 m. The tallest oak in Britain is a mere 43 m (141 ft). But we do have some particularly wide trees, fatter even than General Sherman. The fattest British oak is the Fredsville Oak in Kent, which is only 24 m (79 ft) tall but a huge 3.8 m (12 ft 6 in) wide. Better still, the Fortingall Yew on Tayside, Scotland, although now only represented by two living fragments, had an original diameter that can be traced in the ground of 5.4 m (17 ft 8 in), and a lime was measured by Kannegiesser in Germany in 1906 with a similar diameter. You should try measuring it out to appreciate just how big they are. However, even larger than this is the African baobab (Figure 6.2), which often achieves a diameter of 10 m; Livingstone spoke of a tree in which 20–30 men could lie down with ease! And don't forget the banyan trees mentioned in Chapter 4, which can cover many hectares. You might think that tropical rainforest trees would be up there among the giants. A few are: the tallest tropical tree ever found appears to be a relative of the monkey puzzle (*Araucaria hunsteinii*) in New Guinea at 89 m (292 ft). Surprisingly, though, the tallest tropical trees usually average no more than 46–55 m. Like the coastal redwood, rainforest trees are often remarkably slender for their height. The fattest can be as large as the Fortingall Yew above but diameters greater than 0.3 m (1 ft) are uncommon.

Figure 6.2. Baobab tree (*Adansonia digitata*) in Zimbabwe (photographed by Rosemary Coleshaw).

What limits the size of a tree?

Two linked answers can be given, the first to do with water and the second with investment in mechanical strength.

Tree height usually correlates well with the availability of water. Water shortage is felt more frequently at the top of a tree because it has to be sucked further (see Chapter 3). As water becomes scarcer, water shortage hits at a lower height, and growth at the top slows and eventually stops. The drier the environment the sooner upward growth will be stopped. As an example, coastal redwoods grow taller than 100 m on the deep moist alluvial soils where they are watered by the frequent coastal fogs, but they reach only 30 m on the drier inner edge of the fog belt. As will be discussed below, the final height and shape of any one tree is the product of the environment of the particular spot in which it is growing. However, water stress is not the entire answer to tree height because no matter how much water a tree is given it still has a more or less fixed upper limit: an oak will never grow as tall as a redwood.

The second constraint limiting height is to do with investment in mechanical strength. As a tree gets taller, the wind bending the tree has more leverage on the base in the same way that a longer spanner puts more turning force on a stiff bolt. So, the tree is more likely to snap the taller it is unless, of course, the trunk is made progressively fatter further down. The main problem is that for normal trees height stays in proportion to the cube of the diameter. This means that a tree has to increase the diameter at the base by around eight times to allow for a doubling in height and still remain reasonably safe from breaking when bent by the wind. Obviously, a taller tree has therefore to invest a disproportionately higher amount in wood. This also applies to branches: they have to be disproportionately fatter to resist the bending from (in this case) gravity. A tree needs to be tall to outcompete its neighbours, and big to hold as many leaves as it can, but there comes a point where the return is less than the investment. Evolution has honed the best compromise for each species in each area, producing a genetically determined normal maximum height. You can see that this is genetic because seed from high-altitude and high-latitude trees planted in favourable areas will still produce shorter specimens than seed of the same species from less extreme areas.

As the young tree grows, its width is delicately balanced to be thick enough to withstand the rigours of the environment without being wasteful and overly thick. Once at its maximum height, however, the tree continues getting fatter, producing the magnificent squat hulks so characteristic of the British landscape, where many old trees have been preserved. From what we have said before this would seem mechanically to be a waste. As explained in Chapter 3, however, a tree has no choice but to produce new wood to maintain the water-filled link between the roots and the leaves. And it has added advantage of maintaining a shell of sound wood in the face of rot at the centre of the tree. But this need to produce new wood each year is also partly responsible for the death of the tree (see Chapter 9).

What controls tree growth?

Büsgen and Münch, two German foresters in the early part of the twentieth century, noted that growth in trees depends on 'internal disposition and external influences'. The external influences are obvious enough: growth requires sufficient supplies of light, water, carbon dioxide and nutrients (discussed below), and an amenable climate (see Chapter 9). But the internal disposition is also important. Within a tree there are conflicts over where to use limited resources. Many parts of the tree produce food (carbohydrates) by photosynthesis: mainly the leaves but also the bark, flowers, fruits and buds. Different parts of the tree (described as 'sinks' in the language of physiology) compete strongly for these resources, and a balance has to be kept between the growth of different parts to ensure long-term survival. Too rapid height growth might leave a tree with no stored food for next year's spring leaves. Too many fruits on a young tree might disastrously stunt growth. Fortunately, the tree has a highly organised internal control system for allocating resources to growth, maintenance and reproduction by switching on and off different sinks at different times. The signals for this switching come from the external environment.

Internal control

Without nerves to coordinate activity, a tree relies on hormones (growth regulators and inhibitors), which circulate through the tree to coordinate activities. Three types of growth regulator act to promote various aspects of growth: auxin (indoleacetic acid), produced primarily by shoot tips and leaves (and the basis of commercial rooting compounds); gibberellins, produced in the same place as auxins plus the root tips; and cytokinins, produced particularly by the root tips and young fruit. The gas ethylene should also be included since it appears to regulate wood formation (it is certainly in high concentrations in sapwood and may be involved in responses to bending pressures and other mechanical disturbance, and in heartwood formation) and is involved in fruit ripening (one rotten apple—producing ethylene—spoils the barrel). There are many minor inhibitors in plants but the main one is abscisic acid (ABA), produced in leaves, seeds and other organs. ABA generally slows things down or stops them, causing bud and seed dormancy, and stimulating the shedding of leaves.

Balancing the roots and shoots

This is an important element in the internal disposition or balance of a tree. The shoots provide food for the roots and in turn the roots provide water and minerals to the shoots, and so there needs to be a balance between the two parts (usually referred to as the root : shoot ratio). Too many roots become an undue burden on the limited sugars produced by the canopy; too few roots means water stress for the canopy. Trees have the ability to fine-tune the root : shoot ratio to prevailing conditions of light, water and nutrient availability; this is integrated within the tree by using the hormones described above. For example, if part of the canopy is broken off, some of the roots will die. Conversely, drought leads to a higher proportion of roots. If the roots have problems growing because of shallow soil or competition with other plants, the canopy remains small. The balance of the ratio is at least partly under genetic control: trees from arid areas tend to have higher root : shoot ratios than specimens from moister areas when grown under similar conditions, and tree species that invade open areas tend to have proportionately more roots.

We see the importance of the balance most acutely when it comes to planting trees. A tree dug from a nursery may lose up to 98% of its roots (see Chapter 4). This is why we plant trees in the dormant season to allow some growth of new roots before the canopy comes into leaf and demands water, but is it any wonder that newly planted trees often die of drought? Likewise, young trees planted in containers may be fine for a few years, but as the canopy gets bigger the restricted roots cannot supply enough water to support a large canopy, resulting in small leaves and a tree that remains stunted or dies. If it dies, people often can't understand why: 'It survived that really dry spell a few years ago, why has it died now?'. A corollary of this is that if the roots of a tree are damaged by, for example, cable-laying along urban streets (page 86), the solution to improving the chance of the tree surviving is to reduce the size of the canopy by pruning.

External factors

As noted above, a tree needs optimum supplies of light, water, carbon dioxide and nutrients for optimum growth. A lack of one or more will inevitably slow growth.

Light is often the most crucial factor that comes to mind. Fortunately, trees have a number of ways of countering too little light. A tree growing in shade will be 'drawn up' to the light and will be taller and thinner with shorter branches than a tree in open sunshine (in a paradoxical way, light tends to stunt growth). Trees also vary greatly in their response to shade. Foresters usually express this in terms of shade tolerance. The shade-intolerant (light-demander) species, such as aspen, false acacia (*Robinia pseudoacacia*), larch and many pines, are affected by shade much more than shade-tolerant species, such as beeches, sugar maple (*Acer saccharum*), hemlocks (*Tsuga* spp.), firs (*Abies* spp.), and western red cedar (*Thuja plicata*). Photosynthesis works in the same way in both shade-tolerant and shade-intolerant trees but the shade-tolerant species have more effective mechanisms for coping, such as sun and shade leaves (Chapter 2) and a different canopy 'architecture' (Chapter 7).

Trees, in common with other plants, need nitrogen, phosphorus, potassium (the NPK mixture), calcium, magnesium and sulphur in largish quantities (the 'macronutrients') and small quantities of iron, manganese, boron, zinc, copper, molybdenum and chlorine (the 'micronutrients'). Lack of any of these leads to characteristic symptoms and reduced growth. Plants will go a long way to scavenge back nutrients from being lost; before leaves are shed nutrients are salvaged back into the tree, and it has been observed in teak (*Tectona grandis*), which normally grows on nutrient-poor soils, that nutrients (especially nitrogen) are removed from the body and stalk of the wind-dispersed fruits before being shed (Karmacharya and Singh 1992).

Too little (or too much) water also limits growth. In Chapters 2 and 4 we saw that drought and flooding play equally important roles in reducing the growth of most trees. Indeed, repeated water stress tends to produce a typical shrubby appearance. Although carbon dioxide is not usually limiting to growth—the air always contains a supply—drought can affect its uptake. A water-stressed plant will by necessity limit the opening of the stomata to limit water loss, with the consequence of also limiting carbon dioxide uptake.

The effect of climatic conditions is discussed in Chapter 9, but it is worth mentioning here something about temperature. Foresters, and more lately those interested in climate change, have shown a strong interest in the effect of temperature on growth. Photosynthesis and other physiological processes will occur at temperatures from not far above freezing to around 35 °C. Not surprisingly, the optimum temperature for growth increases as you move from the poles towards the tropics. Whatever the optimal temperature, for many species of hardwoods and conifers, growth is highest under fluctuating temperatures with the nights cooler than the days, probably because of reduced respiration and conservation of food at night. But this is not always the case. Plenty of tropical plants have been found that do better with warmer night-time temperatures (perhaps because warm nights promote leaf production and therefore increase the plant's subsequent capacity for photosynthesis).

Growth in the real world

Trying to work out what is causing a tree to grow slowly can be difficult for a number of reasons.

1. A number of small 'problems', fairly minor by themselves, can amalgamate, leading to a large effect. For example, bonsai trees are kept small by a combination of restricted growth of roots, genetic selection of stock, and sometimes limited water supply. Also, a significant but not disastrous lack of, say, nutrients when coupled with, say, pollution can lead to a very unhealthy tree. And stress of one sort may predispose trees to disease and insect attack because defences are weaker (for example, drought often leads to fungal outbreaks), masking the original cause of stress.

2. The problem may be sporadic and so go unnoticed: harsh temperatures, high winds, drought, etc.

3. Being so big, different parts can be exposed to different conditions. Air temperatures vary more than in the soil so while roots are comfortably cool, the temperature of the cambium in the trunk can be more than 30 °C on the sunny side of a tree and half that on the other side. In the American Midwest, where water is limiting, the north side of a tree grows larger (longer branches and roots) because the south side experiences more water stress. In damp England you expect the opposite! These variations in conditions may not be obvious to a human on the ground.

4. Symptoms may not point directly to the real cause. If a young tree is not growing very well, we automatically reach for the nitrogen-rich fertiliser. And yet in studies done on tree planting in urban areas it has been shown that when nitrogen is applied around trees less than 20 years old the grass benefits far more than the trees! Weed control around the tree is far more effective in stimulating growth than applying fertiliser. (Older and weed-free trees do, however, respond readily to nitrogen.) Competition for nutrients in the soil is a pitched battle that should not be underestimated. We know that certain trees tend to grow naturally on certain soils: in Britain ash is found on nutrient-rich soil (river valley bottoms to dry limestone), birch and pine on nutrient-poor sands and peats, yew on dry chalk and limestone. But if we protect from competition, we can grow all these trees on a wide range of soils. It is also important to realise that the factors limiting establishment from seeds are often different from those limiting subsequent growth. Alders (*Alnus* spp.) can be planted as seedlings on dry soil but wouldn't grow there from seed. On a larger scale, the southern limit of the coniferous forest across Canada is dictated by moisture, the northern limit by summer temperature, but these act primarily on young trees; established trees can persist and grow beyond these limits.

Growth rings and dendrochronology

The interaction of climate with tree growth is well illustrated in the size of tree rings. In favourable years tree grow wider rings than in unfavourable years[1]. Just what does favourable mean? In extreme environments, ring width usually correlates closely with summer temperature whereas in less extreme areas it is either temperature or, more commonly, a mix between temperature and precipitation

[1] Wider and narrower rings are measured in relation to each other. The absolute width of rings varies by species, age of the tree and growing conditions. In any one tree the rings can be bigger on a side that gets more light or due to the presence of reaction wood; rings also tend to be wider at the base of the crown compared with the base of the trunk.

(although in fact it is usually *last* year's weather that the ring responds to; the reason is explained under Value of these strategies below).

There is an important caveat to this. When looking at a cut stump you will almost inevitably see that the rings get narrower towards the outside of the tree. This does not necessarily mean that growing conditions have got worse over the years (despite what our memories tell us about superb childhood summers). If you calculate the area of each year's growth ring on the cut surface ('basal area increment') you will get a measure of the cross-sectional area of new wood added each year, and it tends to be remarkably constant over the years. This is because once the tree canopy has reached a more or less fixed size at maturity, the amount of water needed each year becomes fairly constant, and so the amount of wood grown to carry the water becomes equally constant. If the same cross-sectional area of wood is added each year, as the tree gets fatter the rings must get increasingly narrower (the same amount of wood has to be spread around the outside of the trunk in a thinner layer) until they should almost disappear, and this indeed is when the discontinuous and missing rings described in Chapter 3 commonly appear.

The relationship between climate and ring width has led to the discipline of dendrochronology. Trees of the same species growing in the same area tend to produce similar sequences of wide and narrow rings. On the core taken from tree 1 in Figure 6.3 we can find exactly what year any one ring was growing by counting back from the bark, and if we are dealing with a long-lived species we might get ring sequences many hundreds of years long (what Schweingruber, a Swiss dendrochronologist, has described as a 'tree's private diary'). Now say we find a piece of wood that is dead (core 2 in Figure 6.3) and we have no idea when the wood was growing. If we find ring patterns that match with those somewhere along our first core then we can accurately date the previously unknown piece of wood. By adding new pieces of wood of the same species to the sequence

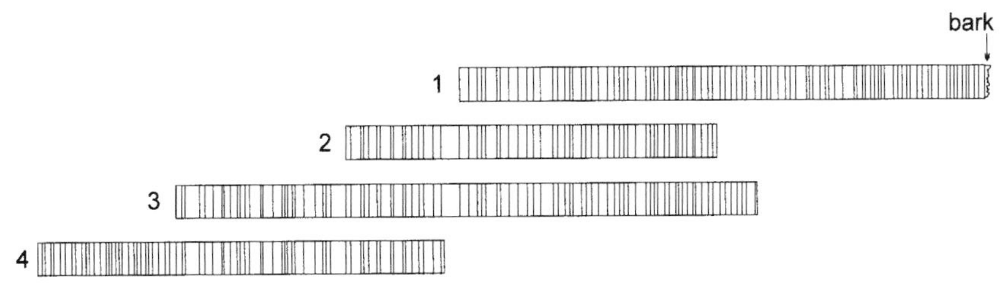

Figure 6.3. Dendrochronology: matching patterns of tree rings from different trees to enable dating of unknown pieces of wood. In a living tree, such as tree 1, the year a particular ring grew can be calculated by counting back from the bark. If the pattern of wide and narrow rings can be matched to pieces of dead wood (2–4), they can also be dated. In this way long 'chronologies' can be built up. (Note that the 'crossmatching' between cores is never quite as perfect as shown here; there is always some variation from tree to tree requiring the use of computers to find the best match.)

we can build up 'chronologies' that extend back sometimes thousands of years. In Britain the longest chronology for oak now stretches back from the present day to over 7000 years ago. This is possible because oak can be found in modern woodlands, historic buildings and older archaeological sites, and is also naturally preserved, thousands of years old, in peat bogs. These chronologies have provided a powerful device for ageing pieces of wood from old buildings, archaeological sites, etc., and for checking the accuracy of carbon dating. There are limitations: a piece of wood from another part of the country, growing under different weather conditions, may be hard to fit against a chronology. But with more sophisticated computer matching programmes, this is becoming more realistic because they can pick out underlying patterns from the local 'noise'.

As well as acting as precise dating tools, tree-ring chronologies can also tell us a lot about past climates. In Scandinavia a chronology for Scots pine has been constructed stretching back 1400 years. The most recent part of this chronology mirrors summer temperature records that have been kept during the historical period, and it has been possible to infer mean summer temperatures for the full 1400 year chronology, giving climate records for a period when none were kept by humans. Thus it is possible to pick out the Little Ice Age (AD 1550–1850) and also the colder and warmer climatic episodes from prehistory.

Buds and tree growth

As we saw in Chapters 1 and 3, trees grow in two ways: there's elongation of the branches (primary growth) and a later fattening of the same branches and trunk (secondary growth). Secondary growth is dealt with in Chapters 3 and 4. Here we will look at the mechanics of primary growth and the important role played by buds.

Buds

In climates with no distinct seasons, growth may be continuous all year round. This is seen, for example, in some of the eucalypts and mangrove species in Florida. But the norm is rhythmic growth with a period of growth followed by a resting stage coinciding with a less favourable time of year. During the resting stage the vulnerable growing tip (the 'meristem') is usually protected from cold, desiccation and insect attack by being enclosed in a bud. As the end of the growing season approaches, the last few leaves (in most plants) are modified into thickened bud scales, which remain on the plant when other leaves fall, enclosing the delicate tip. The defences of the tough, waterproof, overlapping bud scales (officially called cataphylls) are often supplemented by resins, gums and waxes (hence the sticky buds of horse chestnuts). That bud scales are really modified leaves is more obvious when the bud starts growing in spring. Figure 6.4 shows a horse chestnut bud opening and you can see that some of the upper scales bear

(*a*) (*b*)

Figure 6.4. (a) A bursting bud of horse chestnut (*Aesculus hippocastanum*) showing the transition from bud scales through to bud scales with part of a leaf at the tip to proper leaves. This demonstrates that the bud scales are really modified leaves. (b) A few weeks later and the bud scales will soon be dropping off but those with a partial leaf will remain for much longer and act like a small leaf. Keele, England.

leaf blades at their ends, which may stay on the plant longer than the rapidly shed scales. The whole leaf does not have to be used: in the tulip tree (*Liriodendron* spp.) and limes (*Tilia* spp.) the highly toughened stipules (two appendages at the base of a leaf; see Chapter 2) of the upper leaves serve as bud scales.

A number of trees, even temperate ones, don't go to all the fuss of using bud scales. At the end of the growing season the newest leaves simply stop growing and form a 'naked bud' (Figure 6.5). In the spring these young leaves will resume growth as if nothing had happened. Presumably the risk of damage or death during the dormant period is counterbalanced by not wasting energy growing scales that are simply shed in spring. Still, eucalypts that have naked buds have an insurance policy: the main buds are accompanied by smaller 'concealed' buds covered by the leaf base. It must also be said that some conifers, especially in the cypress family, including the false cypresses (*Chamaecyparis* spp.), junipers (*Juniperus* spp.) and thujas, do not have distinct buds at all; instead they produce new growth from meristematic tissue hidden under the skin of the twig.

Figure 6.5. The naked buds—with no bud scales—of the wayfaring tree (*Viburnum lantana*). At the end of the growing season the newest leaves simply stop growing and form a 'naked bud'. The leaves resume growth in the spring where they left off. From: Oliver, F.W. (1902). *The Natural History of Plants.* Vol. 1: *Biology and Reproduction of Plants.* Gresham, London.

The new shoot

Dissect a bud and you find next year's shoot preformed in miniature. The twig and leaves (and maybe flowers) are there waiting. So when spring comes and the bud bursts open, the new shoot can grow out quickly because it is just a case of expanding what is already there, in the same way that inflating a balloon is quicker than starting with a bowl of rubber solution and having to make the balloon first.

Buds of different trees vary in just how much of the next year's growth is preformed. In fact buds can be divided into three types of growth. In trees such as ashes, beech, hornbeams, oaks, hickories, walnuts, horse chestnuts and many maples and conifers, the whole of next year's shoot is preformed. Everything is preformed or fixed in the bud as it develops over the summer to be expanded into a branch the following spring. These species are described as showing **fixed or determinate growth**. Since everything is preformed, spring growth occurs in a single, rapid flush and is all over in just ten days to a few weeks (Figure 6.6), and then the terminal bud immediately takes on its winter appearance.

In many others, however, only some of the leaves are preformed. Once these

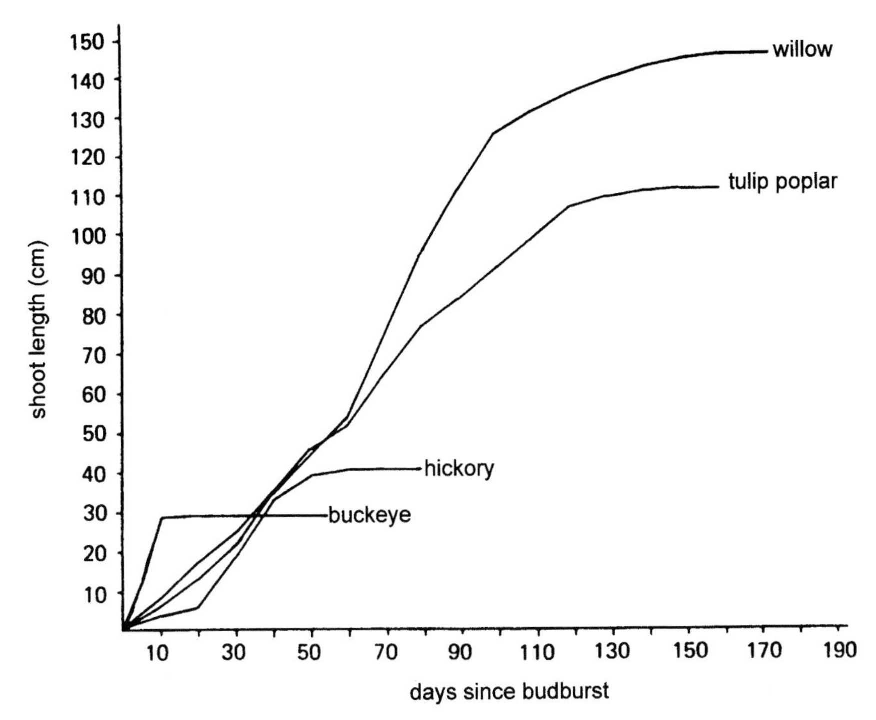

Figure 6.6 . The growth of shoots of a number of trees in Georgia, USA. Measurements of shoot length were made every two weeks on nine trees ranging from 8 to 15 years old. Buckeye (*Aesculus georgiana*) and hickory (*Carya tomentosa*) stop growing in a few weeks ('fixed growth') whereas the tulip poplar (*Liriodendron tulipifera*) and willow (*Salix nigra*) continue growth for a greater part of the summer ('free growth'). Modified from: Zimmermann, M.H. and Brown, C.L. (1971). *Trees: Structure and Function.* Springer, Berlin, Figure I-19, Page 45.

have expanded, given suitable weather, the shoot will continue to produce other leaves from scratch. As an example, the tulip tree (*Liriodendron tulipifera*) usually produces 14–20 leaves on a shoot over a summer, of which eight are normally preformed. The 'early' preformed leaves are sometimes a different shape from the 'late' leaves (called heterophylly) so you can see just how many were preformed; early leaves may be less deeply lobed or divided as in sweetgum and ginkgo, or proportionately wider as in silver birch. Trees showing this **free or indeterminate growth** (sometimes also called **continuous growth**) include elms, limes (lindens), cherries, birches, poplars, willows, sweetgum, alders, apples, and conifers such as larches, junipers, western red cedar, the coastal redwood and ginkgo. Many of these, you will notice, are early invaders into open areas. Growth continues for longer than in determinate species but still normally stops well before the end of the growing season, giving time for next year's early leaves to be preformed in the new buds.

Both fixed and free type trees will sometimes show a second burst of 'lammas' growth from the terminal buds (called lammas because it usually occurs around Lammas Day: the 'bread-feast' harvest festival, traditionally 1 August or the seventh Sunday after Trinity). This is typical of young oaks but is also seen in elms, hickories, beech, alder and a number of conifers including Scots pine, Douglas fir and Sitka spruce. In oaks, lammas growth can be identified by the more deeply incised leaves; in pines there may be more needles per bunch than normal, but shorter, giving a tufted appearance.

The third strategy is intermediate, growing by **recurrent flushes (rhythmic growth)** such that there are several cycles of growth and bud formation in a year. This is commonly found in the fast-growing southern pines from the warm south and east USA—e.g. loblolly (*Pinus taeda*), shortleaf (*P. echinata*) and Monterey pines (*P. radiata*)—and the Caribbean (Caribbean pine, *Pinus caribaea*), and more tropical species such as cocoa (*Theobroma cacao*), rubber tree (*Hevea brasiliensis*), avocado (*Persea americana*), mango (*Mangifera indica*), tea (*Camellia sinensis*), lychee (*Litchi chinensis*) and citrus species, plus the European olive (*Olea europaea*). Tea, for example, generally has four flushes of leaves per year (just as well for the tea producer since it is the new shoot tips that are used!). Certainly in tea and rubber, these flushes do not seem to be induced in response to the external environment and so must be internally controlled. Sometimes the recurrent flushing can merge towards continuous free growth. When this happens in pines it results in a long branchless leader with a tuft of needles at the end, aptly called 'foxtail' growth.

Value of these strategies

Trees with fixed growth, producing all their leaves in one go in spring, would seem to be under a distinct disadvantage. Firstly, it makes them less responsive to good growing conditions in the current year; the amount of growth this year is correlated to the weather of the previous summer when the shoots and leaves were preformed in the buds. This could mean that the benefits of a good summer will be diminished if the tree is handicapped by having small buds with few preformed leaves as a result of a previous bad summer. The tree might compensate by growing longer shoots and making its few leaves larger but it is unlikely to make up completely.

So where does the advantage of preforming all next year's growth lie? The answer takes us back to the differences between ring-porous and diffuse-porous wood discussed in Chapter 4. Ring-porous trees (such as oak and ash) have to grow a new ring of wood before they can produce leaves, which means they generally leaf out later than diffuse-porous trees. Lo and behold, most ring-porous trees make up for this by having fixed growth, which means they produce a full set of leaves without any further delay. The advantages of not losing any more time must outweigh the disadvantages. But this is probably not the whole answer because elm, a ring-porous tree, has free growth, and diffuse-porous trees like limes, planes, tupelos (*Nyssa* spp.), magnolias and the sweet gum (*Liquidambar styraciflua*) leaf out even later, apparently wasting some of the growing season. This late leafing habit could be because these trees were part of the tropical and semitropical flora in the Tertiary (spread over the past 65 million years) compared to the more Arctic connections of the earlier-leafing diffuse-porous trees.

This still begs the question of why show fixed growth. It makes sense to produce a full set of leaves rapidly but why not tack on a few more later in the summer if the conditions are good? Why stop growth of the branches so early in the summer when growing conditions are just reaching their best and extra valuable height could be gained by continuing a little longer? Even in trees with free growth, branches normally stop growing well before the end of the growing season. Why have this period of 'summer dormancy'?

The answer, unfortunately, is not clear but may be to do with maintaining sufficient food reserves within the tree. The spring growth in deciduous trees may sufficiently deplete the food reserves that further growth is prevented until they are replenished. This doesn't mean that spring growth completely drains *all* the reserves of a tree: some reserves are kept for emergencies so that, for example, a tree defoliated in spring can grow a second flush of leaves. But it is probably sufficiently fine a balance (like aiming to keep a bank balance just in the black despite emergencies) that once the new branches (and roots) are grown, and energy is put into maintaining the tree, flowering, fruiting and replenishing the food reserves, there is little left over for further branch growth that year except under exceptionally good conditions or severe defoliation (such as lammas growth; see previous page). Bear in mind that in large old trees (where more than 90% of the mass may be in the woody skeleton), respiration of the living tissue takes one to two thirds of the carbohydrate produced (tropical trees tend to be at the higher end). This may explain why young trees (with a lower respiratory burden) and coppice shoots (with abundant food stored in the roots) show growth for longer in a season than a mature tree.

I am sure this is still not the whole story. For one thing, studies have shown that reserves are rapidly replaced and are normally complete by around late July in north temperate areas. Also, evergreen species, which finance new growth from the photosynthesis of old needles rather than from reserves, still show a halt in growth early in the season. Alternative arguments can be built around the avoidance of waste. If new outer leaves on vigorous, never-ending growth start shading those produced earlier in the year then the tree is wasting resources on inferior leaves. Moreover, there will come a point in the summer where new leaves will cost more to grow than they can hope to recover in photosynthesis in the short time left in the season. Indeed, those trees with repeated flushes of leaves are found in areas of long or continuous growing seasons and many are either typical of forest gaps (and so less likely to suffer from self-shading) or regularly shed their leaves.

Last of all, reduced root growth in summer (and therefore reduced shoot growth to keep roots and shoots in balance) and mid-summer water shortage at the top of mature trees has also been implicated in the argument, which helps add to the complexity and confusion! Like many complex questions, the truth undoubtedly lies somewhere in a mixture of these possible explanations.

Phenology: timing and pattern of annual growth

Trees have inherent cycles of growth. In the tropics these may be in tune with the tree's own internal rhythm but elsewhere the cycle of growth is regulated by seasonal changes. In such trees the link between the strategic use of resources and the seasonal changes is one of great finesse.

The first part of the tree to stir in spring, several weeks before bud break, is the roots (although if the winter was mild they might never have stopped). This makes good sense because a high priority in spring is water to expand the new shoots. Roots don't do anything special to survive winter, they merely stop growing until soil temperature gets above *c.* 5 °C; see page 101. Once the buds break, root growth slows down (Figure 6.7). This may partly be due to limited resources

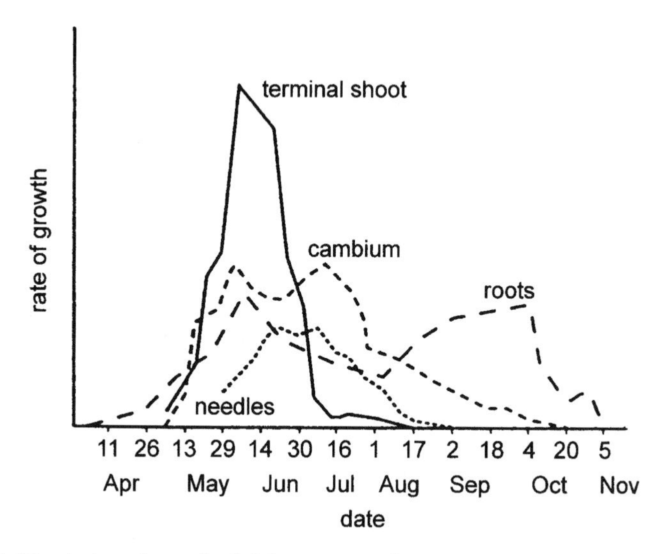

Figure 6.7. The timing of growth of different parts of a 10 year old eastern white pine or Weymouth pine (*Pinus strobus*) in New Hampshire, USA. The roots begin elongating before the shoots and show a mid-summer dip in growth. The needle length and trunk diameter (cambium growth) keep increasing over much of the summer but around 90% of all shoot growth (including height) occurs in just 6–9 weeks early in the season. Modified from Kienholz, R. (1934). Leader, needle, cambial and root growth of certain conifers and their relationships. *Botanical Gazette*, **96**, 73–92. © 1934 by the University of Chicago. All rights reserved.

being commandeered by the active above-ground growth above or due to soils getting a little too warm or dry for root growth. Either way, root growth speeds up again in the autumn.

Activity above ground starts with the breaking of the buds. For trees like beech and rhododendron, the increasing day length of spring triggers bud burst (though strictly speaking the plant measures night length through a pigment called phytochrome found in buds and leaves). For the majority, however, bud burst is triggered primarily by temperature. Just how warm it needs to be depends not just on species but also upon geographical location (trees get going at lower temperatures in cold areas) and how cold the winter has been (which affects winter dormancy; see below). In any one area, different species tend to leaf out in the same order each year, spread over 2–3 months, but the whole sequence may be shifted by several weeks depending on the warmth of the spring. Having said that, there is often some small variation in sequence since different species respond to different combinations of conditions; hence the old rain predictor 'oak before ash, in for a splash; ash before oak, in for a soak'. Tropical trees may appear to use equally prominent environmental signals: in dry tropical forests new leaves appear on the bare trees at the start of the rainy season. But the buds break just *before* the rains, and are really induced by slight changes in temperature.

Once the buds begin to open, a hormonal signal is passed to the cambium to start producing new wood (xylem) and inner bark (phloem). In diffuse-porous trees (like beech and birch) and conifers, phloem is produced right away, ensuring the mobilisation of stored food to the growing points. New wood is not such a high priority in these trees because the wood from previous years is quite capable of conducting enough water before the new leaves are in full swing. So new wood growth starts at the buds and progresses down to the base of the tree over several weeks and reaches the end of the roots in another 4–6 weeks. The same is not true for ring-porous trees such as oak and elm. As we saw in Chapter 3, these trees move the bulk of their water through the ring of wood formed the same spring. These trees begin new wood formation rapidly through the whole tree, up to 2–3 weeks before bud break. Cambium growth is stimulated by hormones (auxins) from developing buds, so how can the trees start so early? The answer is that the cambium is sensitive to the very low concentrations of auxins produced by the buds just beginning their development. Xylem growth in most trees may go on till mid-summer, tapering off into autumn (Figure 6.7). When the cambium is most active it is a physically weak tissue and the bark is most easily peeled from the tree; the season for peeling oak bark for tanning is reckoned to be from May to June. (Incidentally, if you want the bark to stay on a piece of wood for ornamental purposes, don't cut the tree during this period; if you do it won't be long before the bark simply falls off.) Phloem production continues for a shorter time than xylem because there is less of it produced.

As the season warms, the buds on a tree progressively open in a more or less regular pattern starting at the terminal bud and working back along the branch.

Traditionally it is said that bud expansion starts at the top of the crown but it is my experience that, in woodlands at least, leaves often appear first on the lower branches. Generally, leaves and shoots grow more at night when there is less water stress. There are, of course, plenty of exceptions to this in both temperate and tropical trees for a variety of reasons. For example, in Sweden the leaves of the common osier used for basket weaving (*Salix viminalis*) have been found to grow more during the late afternoon and early evening because of cold nights.

Trees are masters of forward planning: as the current set of buds open and the new shoots grow out, next year's buds are already forming! If the tree is old enough, some of these buds will contain flowers. In temperate hardwoods and conifers, flower buds are formed in late summer or autumn, opening the following spring or summer. There are (of course) exceptions: some warm-temperate trees (e.g. buddleia, fuchsia and hibiscus) form their buds in spring just prior to flowering and, at the other end of the scale, eucalypts start their flowers two or more years before opening. In the more equitable climate of the tropics, flower buds are often initiated more evenly over the year, taking just a few months to fully form, and these may open as they form (as in fig species) or be accumulated until environmental conditions are right for them to open, which may be sometimes three times a year (as in strangler figs) or only once every 5–6 years as in the rainforests of Indonesia. Generally the better the growing season when the buds are formed, the greater the number of flowers. Paradoxically, imposing a stress on trees, such as drought, can also stimulate a 'stress crop' almost as if the tree is making a valiant effort to produce a crop of seeds in case it is its last! Ring-barking of the trunk has also been used to induce flowering in fruit trees and conifers; in this case the idea is to keep more of the carbohydrates produced by the leaves in the canopy, preventing them being 'wasted' on the roots. This is a dangerous operation, which can kill the tree if taken too far.

The trigger for flower buds to open may be different from that for leaf buds, explaining why flowers can open either before or after the leaves (Chapter 5). Woody plants in general seem less sensitive to day length than most herbaceous plants, but there are a few trees, especially in the tropics, that are stimulated to flower by short days (e.g. tea, coffee, bougainvillaea and poinsettia). The most general trigger of flower opening seems to be temperature. In Canada this is being exploited by Elisabeth Beaubien who, with the help of schools, is drawing up maps of flowering times of different species across the prairies and correlating these with temperature. Her aim is to help farmers correctly time their farming operations and predict harvest time by using these wild thermometers as indicators. The required temperature changes can be quite subtle; the mass flowering of many trees together in the tropical forests of Borneo and Malaysia (Figure 5.11) appears to be triggered by a decrease in minimum night-time temperatures of just 2 °C for three or more nights. In addition to temperature, evolution has equipped plants with a range of specialist triggers to ensure they flower and fruit at the best time. Triggers may be rainfall in arid areas, fire for some eucalypts, or even, as in one particular Central Australian shrub, complex interactions between time of year, rainfall, insects and birds.

This brings us to the end of the growing season, the shedding of leaves and the start of dormancy for the buds. A few trees like the tulip trees and the honey-locust (*Gleditsia triacanthos*) will keep growing until stopped by the cold days of autumn. Most trees, however, detect the shortening days (again by using phytochrome) and/or lower temperatures and make preparations. In deciduous trees, useful minerals are removed from leaves and safely stored before the leaves are shed (Chapter 2). Up till this point, buds formed during the year have not been truly dormant; rather they have been suppressed by the leaves around them. If the leaves are removed or damaged, or growing conditions are bolstered by high rain or fertiliser, the buds will grow at once. Now, however, there comes a period of internally imposed dormancy where even if favourable conditions are given they will still not grow. Dormancy is most commonly broken by exposure to low temperatures (around 5 °C) for something around 300 hours in Britain. This dormant period ensures that the trees don't start growth during a warm autumn or a mild mid-winter spell. Once the dormancy requirements have been met, the tree is once again able to respond to favourable conditions. Wilson (1984) quotes the German custom of collecting cherry twigs on St Barbara's Day (4 December) so that they will flower at Christmas; if they are collected in November they won't flower. For the same reason potted Christmas trees need care. By the time we bring them indoors they may have completed their dormancy requirements and start growing, which won't do them any good when taken back out in the new year (it is best to keep them in a garage for a month or so to convince them that winter is coming back again).

Tinkering with phenology

Phenology can cause us problems. Northern trees moved into southern gardens can drop their leaves annoyingly early, and southern trees taken north will continue growth later in the season than is often good for them. Trees growing near street lights will perceive long days late into the autumn and often keep their leaves long after others around have shed theirs. Fortunately, most trees, even when given long days and moderate temperatures, will eventually go into autumn of their own accord.

If owing to global warming the British winters keep getting shorter and warmer, as seems to be happening, we can expect additional changes in spring phenology. Those trees like hawthorn (*Crataegus monogyna*), whose chilling requirements are easily met, should respond to the warmer spring by opening their buds earlier. On the other hand, trees like beech (*Fagus sylvatica*), which currently only just receive enough chilling in British winters, will be left needing more winter cold. They will still open their buds but only after they have received a long period of spring warmth (almost as if the trees need extra convincing that winter really has finished). In this case climatic warming will *not* bring about earlier budburst and, in fact, will delay it.

Lifetime changes in growth

As I sit exhausted and the children come and ask me to play games *again*, I wonder where they get their energy from. If trees were capable of thought, they might well think the same of their offspring. Trees go through a juvenile period in which reproduction, while not impossible, is rare, the rooting of cuttings is easy and, most importantly here, growth is rapid and continues for long periods. But inevitably, as the trees mature they slow the pace to reach quiet middle age. In deciduous trees like oak, the juveniles show free growth and grow right through the summer, but this gradually reduces to the few short weeks of 'fixed growth' seen in adults. Juvenile tropical trees may shoot up vertically for several years before slowing down, branching and changing from evergreen to deciduous. Trees like tea, which flush several times in a year, slow from the youthful exuberance of 8–10 flushes each year to the sedate normal of four. Leaf shape, shade tolerance and the whole shape of the canopy can change with maturation (see Chapters 2 and 7, respectively).

If you plot the height of a tree over its life against its age, it will form an S-shaped curve (Figure 6.8). That is, growth is initially slow in the seedling, increasing to a maximum at the end of the juvenile phase when it has a good-sized canopy but only a small amount of woody and other tissue to keep alive. As the tree gets larger with a bigger woody skeleton (Figure 6.9)—which takes more energy to maintain (see Chapter 9)—and as it reaches its maximum height determined by genetics and water supply (see the beginning of the chapter), the height and mass gains level off.

The start of reproduction

The final change from juvenile to mature adult is normally marked by the onset of flowering (although some trees will delay first flowering for many years after they are 'vegetatively mature'). Pioneer trees that invade open areas and so need to produce seeds quickly before they are squeezed out will flower early, usually within their first decade (Box 6.1) although, as an extreme, teak has been seen to flower at just 3 months under optimum nursery conditions. By contrast, long-lived trees that invade at a later stage put their resources into dominating before looking towards the next generation and flowering. Thus beech and oak in Britain are *normally* around 60 years old before they flower. Before you start writing me letters about precocious trees you know of, I should point out that many factors affect age of first flowering. The better the growing conditions the younger a tree will flower. Shade is a primary factor. Oaks in the open can begin flowering at

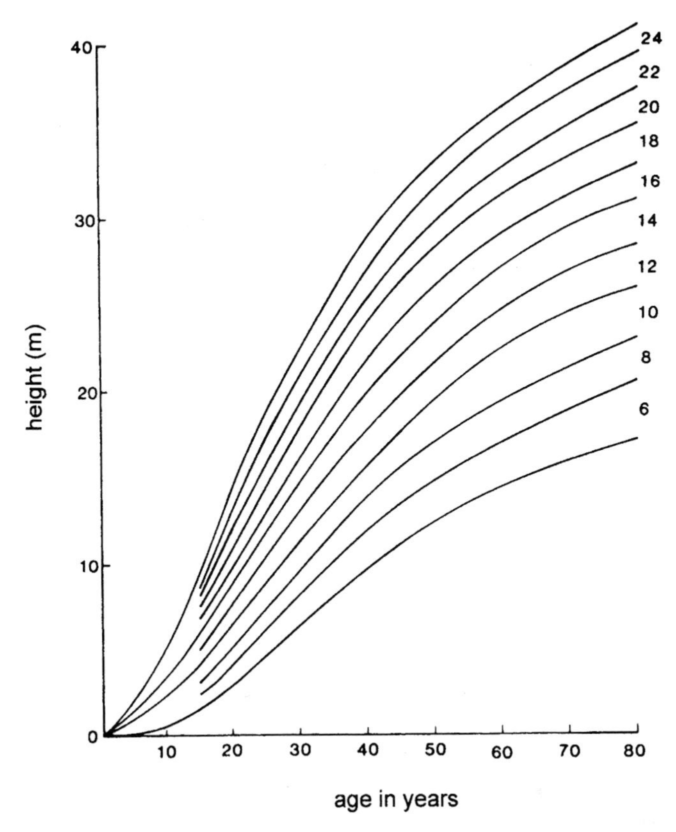

Figure 6.8. Height growth with age for Sitka spruce (*Picea sitchensis*) under different growing conditions. The numbers are the number of cubic metres of wood that are added to a hectare of trees per year (refered to as 'yield classes'). The higher the number, the better the growing conditions. Note that however fast the trees grow, the curve is still the an S-shape. From: Hibbard, B.G. (1991). *Forestry Practice.* Forestry Commission, Handbook 6. HMSO, London, Figure 11.2, Page 152.

less than 40 years old but in heavy shade they may be approaching a century. Male flowers are often produced first because they are cheaper to produce when the tree is still putting most of its resources into growing (see Chapter 5).

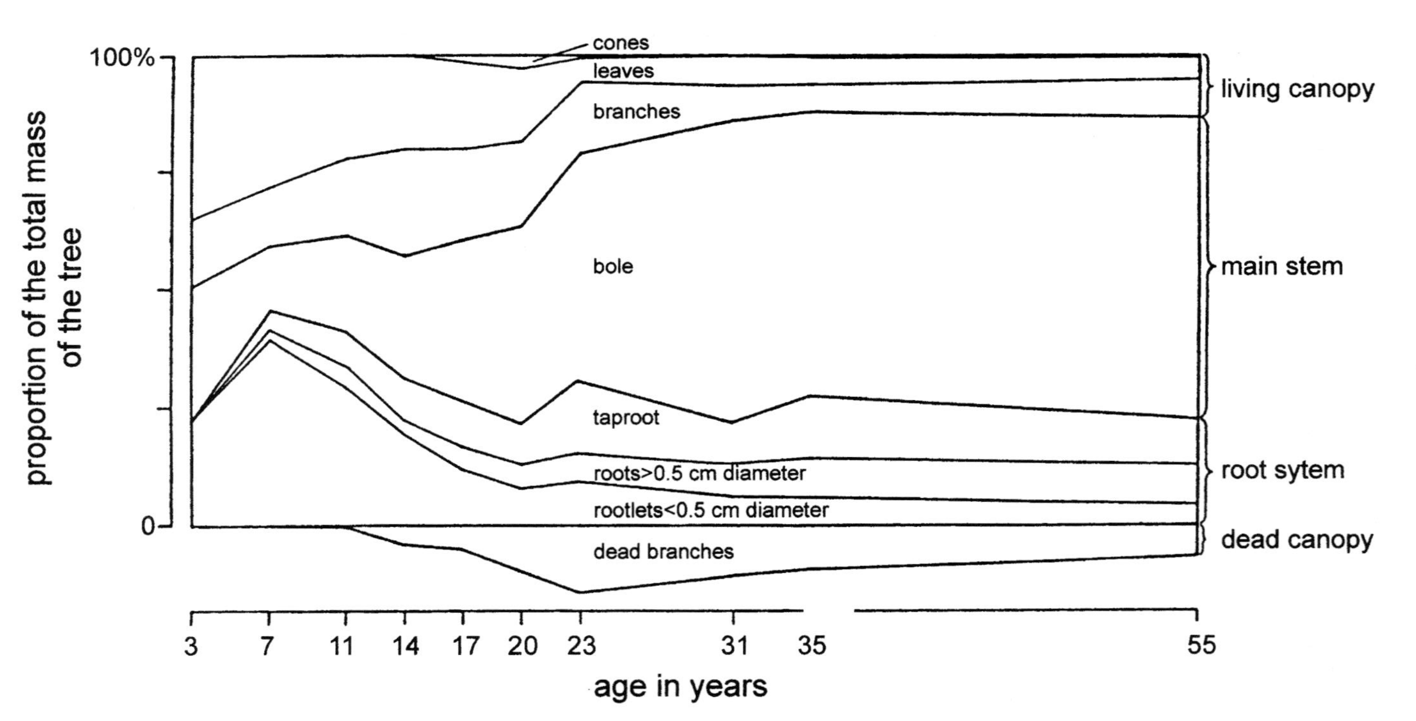

Figure 6.9. Changes in proportion of mass of various parts of a Scots pine (*Pinus sylvestris*) as the tree ages. From: Ovington, J.D. (1957). Dry matter production by *Pinus sylvestris* L. *Annals of Botany (London)* [N.S.] 21, 287–314.

Box 6.1. Approximate age of first fruiting and optimum seed-bearing

Actual age in any one tree may vary either way of the figures given here depending upon growing conditions.

Common name	Latin name	Age of first seed production	Optimum seed-bearing age
Eucalypts (in USA)	*Eucalyptus*	1–3	—
Common walnut	*Juglans regia*	4–6	30–130
Poplars, willows	*Populus, Salix*	5–10	25–75
Coastal redwood	*Sequoia sempervirens*	5–15	250 onwards
False acacia	*Robinia pseudoacacia*	6	15–40
Hazel	*Corylus avellana*	10	—
Honey-locust	*Gleditsia triacanthos*	10	25–75
Giant sequoia	*Sequoiadendron giganteum*	10	150–200
Birches	*Betula*	10–40	20–70

Box 6.1. *(cont.)*

Common name	Latin name	Age of first seed production	Optimum seed-bearing age
Alder	*Alnus*	12	—
Pines (normal values)	*Pinus*	10–60	50+
Monterey pine	*P. radiata*	5–10	15–20 onwards
Ponderosa pine	*P. ponderosa*	7	60–160
Scots pine	*P. sylvestris*	10–15	40–80
Pinyon pine	*P. monophylla*	35	100
Spruces (normal values)	*Picea*	10–40	—
White spruce	*P. glauca*	(4) 10–15	60+
Black spruce	*P. mariana*	10	100–200
Engelmann spruce	*P. engelmannii*	15–40	150–250

Box 6.1. *(cont.)*

Common name	Latin name	Age of first seed production	Optimum seed-bearing age
American elm	*Ulmus americana*	15	40 onwards
Tulip tree	*Liriodendron tulipifera*	15–20	up to 200
Larches	*Larix*	15–25	40–400
Horse chestnut	*Aesculus hippocastanum*	20	30
Western red cedar	*Thuja plicata*	20–25	40–60
Firs	*Abies*	20–45	40–200
Hornbeam	*Carpinus betulus*	20–30	—
Sweetgum	*Liquidambar styraciflua*	20–30	up to 150
Ash	*Fraxinus excelsior*	20–40	40–100
Sycamore	*Acer pseudoplatanus*	20–40	40–100

Box 6.1. *(cont.)*

Common name	Latin name	Age of first seed production	Optimum seed-bearing age
Lime	*Tilia*	25	50–150
Field maple	*Acer campestre*	25	50–150
Red oak	*Quercus rubra*	25	50 onwards
Douglas fir	*Pseudotsuga menziesii*	30–35	50–60
Wych elm	*Ulmus glabra*	30–40	40
Sweet chestnut	*Castanea sativa*	30–40	50 onwards
Hickories	*Carya ovata, C. laciniosa*	40	60–200
Beeches	*Fagus*	40–60	80–200
English oak	*Quercus robur*	40–60[a]	80–120 onwards

Box 6.1. *(cont.)*

Common name	Latin name	Age of first seed production	Optimum seed-bearing age
Tropical trees			
Pioneer		2–5	—
Teak	*Tectona grandis*	5–15	—
the majority		10–30	—
Dipterocarps	*Dipterocarpus*	60	—

[a] Coppice shoots may begin fruiting at 20–25 years of age.

Based on Skene, M. (1927). *Trees.* Thornton Butterworth, London; Hibberd, B.G. (1986). *Forestry Practice.* Forestry Commission Bulletin 14. HMSO, London; and Burns, R.M. and Honkala, B.H. (1990). *Silvics of North America.* Vol. I, *Conifers;* Vol. 2, *Hardwoods.* USDA Forest Service, Agricultural Handbook 654.

Further reading

Beaubien, E.G. and Johnson, D.L. (1994). Flowering plant phenology and weather in Alberta, Canada. *International Journal of Biometeorology*, 38, 23–7.

Cannell, M.G.R. and Dewar, R.C. (1994). Carbon allocation in trees: a review of concepts for modelling. *Advances in Ecological Research*, 25, 59–104.

Friedel, M.H., Nelson, D.J., Sparrow, A.D., Kinloch, J.E. and Maconochie, J.R. (1993). What induces central Australian arid zone trees and shrubs to flower and fruit. *Australian Journal of Botany*, 41, 307–19.

Gower, S.T., McMurtrie, R.E. and Murty, D. (1996). Aboveground net primary production decline with stand age: potential causes. *Trends in Ecology and Evolution*, 11, 378–82.

Hodge, S.J. (1991). Improving the growth of established amenity trees: fertilizer and weed control. *Arboriculture Research Note* 103–91.

Karmacharya, S.B. and Singh, K.P. (1992). Production and nutrient dynamics of reproductive components of teak trees in the dry tropics. *Tree Physiology*, **11**, 357–68.

Kelly, P.E., Cook, E.R. and Larson, D.W. (1992). Constrained growth, cambial mortality, and dendrochronology of ancient *Thuja occidentalis* on cliffs of the Niagara escarpment: an eastern version of bristlecone pine? *International Journal of Plant Science*, **153**, 117–27.

Kozlowski, T.T. (1992). Carbohydrate sources and sinks in woody plants. *Botanical Review*, **58**, 107–222.

Lechowicz, M.J. (1984). Why do temperate deciduous trees leaf out at different times? *American Naturalist*, **124**, 821–42.

Lescop-Sinclair, K. and Payette, S. (1995). Recent advance of the arctic treeline along the eastern coast of Hudson bay. *Journal of Ecology*, **83**, 929–36.

McDonald, A.J.S., Stadenberg, I. and Sands, R. (1992). Diurnal variation in extension growth of leaves of *Salix viminalis*. *Tree Physiology*, **11**, 123–32.

Murray, M.B., Cannell, M.G.R. and Smith, R.I. (1989). Date of budburst of fifteen tree species in Britain following climatic warming. *Journal of Applied Ecology*, **26**, 693–700.

Power, S.A. (1994). Temporal trends in twig growth of *Fagus sylvatica* L. and their relationship with environmental factors. *Forestry*, **67**, 13–30.

von der Heide-Spravka, K.G. and Watson, G.W. (1990). Directional variation in growth of trees. *Journal of Arboriculture*, **16**, 169–73.

Wilson, B.F. (1984). *The Growing Tree*. University of Massachusetts Press, Amherst.

Chapter 7: The shape of trees

The whole point of a woody skeleton is ultimately to get the leaves above competitors to ensure a lion's share of the light. And from this simple goal comes an enormous range of tree shapes, from the unbranched stems of palms and tree ferns to the tall spires of conifers, the broad spreading crown of oaks and the multiple stems of an old yew. What governs the shape of trees? How are trees organised to display what often looks like an impossibly large number of leaves?

Trees of distinctive shape

It is usually possible (but not always!) to identify a conifer from a distance by its conical outline. Within the cone there are usually plates of foliage showing where the branches are produced in whorls around the main central stem, usually one whorl per year (Figure 7.1). This contrasts with the rounded dome of a hardwood where the initially leading shoot of the young tree gives way to a number of strong branches, giving the whole canopy a rounded shape.

Within these two main shapes it is possible (with a little practice) to distinguish different species simply by their shape. This book is not the place to list the distinctive features of common species but one example will illustrate the point. In common lime (*Tilia × europaea*) the main branches develop in great arching curves, which in time lose the terminal buds. New growth comes from near the branch end, resulting in another arch, creating the effect of multiple rainbows joined together at their ends (Figure 7.2). Epicormic buds (Chapter 3) characteristically produce a mass of sprouts around the base of the trunk and frequently a congested growth of small twigs in the centre of the canopy. Some of these young growths escape to produce vertical branches through the crown, parallel to the main stem.

Why have these distinctive shapes?

Shape is a compromise between displaying leaves without undue self-shading, the needs of pollination and seed dispersal, optimum investment in the woody structure ('biomechanics') and coping with the surrounding environment (high winds, poor soil, etc.). In the tropics, where you would expect most variation in

Figure 7.1. Sitka spruce (*Picea sitchensis*) with the typical spire of conifers with whorls of short branches. Perthshire, Scotland.

shape because of the huge number of species[1], trees have been categorised into just 25 or so shapes (see Hallé *et al.* 1978). This suggests that there are a limited number of shapes that produce a workable compromise.

The nature of the compromise is often hard to pinpoint but some gross generalisations are possible. Conifers of high latitude and altitude are typically steeply pyramidal with short often downward-sloping branches to help shed snow (Figure 7.1); this shape also helps intercept the maximum amount of light from the sun low on the horizon (but how do they cope when in dense stands?). Conifers on dry sites have a similar shape but in this case it appears to be an adaptation to intercepting the *least* light (and therefore heat) at noon when it is hottest, saving on the need for cooling by water loss. By contrast, the broad crowns of most hardwoods and some spruces and firs are associated with moist sites, deep shade (a wide canopy intercepts maximum light) or harsh tree-line conditions (to keep low and out of the wind). It is also a good shape for inter-

[1] A hectare (2.47 acres) of tropical rainforest usually contains 60–150 different species and occasionally up to 300. By contrast temperate forest averages around 25–30 species in a hectare. The temperate forests of N America contain fewer than 400 species of tree whereas Madagascar alone has 2000!

(*a*) (*b*)

Figure 7.2. The distinctive shape of common lime (*Tilia* × *europaea*). (a) The typical arching curves of the main branches; because the end buds tend to die, new growth comes from near the branch end, resulting in another arch, creating the effect of multiple rainbows joined together at their ends. The tree in (b) has the characteristic congested growth of small twigs in the centre of the canopy. What you can't see are the mass of shoots that normally spring from the base of the tree. Staffordshire, England.

cepting most light in cloudy climates where the light is diffuse and in effect comes from over the whole sky. Because in Britain we have an almost uniformly moist (and mostly cloudy) environment, most of our trees take on the same, roughly spherical, shape. Pines further south in savanna-type climates, such as the Mediterranean stone pine (*Pinus pinea*), develop a flat-topped, umbrella shape (Figure 7.3), which helps resist drying winds (the aerofoil shape of the canopy allows leaves to hide behind each other out of the wind) and maximises convective heat loss by allowing free passage of air up through the canopy.

The dynamic tree: reacting to the changing world

Having said that a species has a characteristic shape adapted to the type of environment, it is obvious that the shape of an individual tree will vary with conditions. Light, wind, snow, herbivory, fire, root health and many other factors can all influence shape. Trees grown in the open, for example, have a bushy crown

Figure 7.3. Stone pine (*Pinus pinea*) growing in Tuscany, Italy, showing the aerofoil shape of the canopy, which helps in conserving water.

with wide spreading branches while those in shade are taller and narrower with fewer side branches, in an effort to reach light. Mountains provide ample evidence of the effect of climate on tree shape.

Puffing up a mountain you notice that, as a result of lower year-round temperature and increased wind, conifers become shorter and more squat. Quite literally, shaking (or even rubbing) a tree stunts its growth: a glasshouse study by Neel and Harris in 1971 showed that shaking sweet gum trees (*Liquidambar styraciflua*) for just 30 seconds a day reduced the height growth to less than a third that of the unshaken controls, partly because the shaken trees stopped growth and produced terminal buds. This makes good mechanical sense: the stronger the wind the more leverage is exerted on the tree so the optimum design will be more squat (see the discussion on tree height and engineering in Chapter 6).

Higher up the mountain, trees only survive in tight, isolated clumps usually called Krummholz (German for 'crooked wood') although purists will sometimes use the word Kruppelholz. Harsh winds carrying ice particles in the winter wear away the waxy coating on needles like sandpaper, leaving them open to death by dehydration. This often produces 'banner' or 'flagged' trees looking like a flag blowing in the wind with branches surviving only in the lee of the prevailing wind (Figure 7.4). Similar flagging can be seen in other windy places such as on cliffs by the sea where the wind is aided in its damage not by ice but by salt spray. New trees can only establish in the shelter of others (often by layering—rooting—of the branches), leading to clumps of stems huddled together. As whole stems on the windward side are slowly killed and new ones grow on the lee side, the whole clump moves downwind. In N America this has been estimated at 2–7 m per century. At the bottom of these clumps, healthy branches survive like

(*a*)

(*b*)

Figure 7.4. Subalpine fir (*Abies lasiocarpa*) in the Canadian Rocky Mountains demonstrating the effect of hostile conditions at high altitude. (a) A 'flagged' specimen, bare on the side facing prevailing winds; (b) a larger clump of 'Krummholz' with the prevailing winds moving into the picture. Note the skirt of healthy foliage around the bottom showing where snow protects against winter winds.

a thick prickly skirt: these are protected from the ravages of the wind by snow lying on the ground. Higher still up the mountain and the skirt is the only part to survive and the noble spire of lower altitudes has been reduced to a prostrate shrub by environmental conditions.

But trees are not passive in the face of changes to their shape made by the environment: they react. If branches are lost, new ones can be grown from stored buds or new adventitious ones (Chapter 3). Alternatively, an existing branch can be used to fill the gap, perhaps by reorientation through bending (see reaction wood, below) or by giving it a more major role. This is seen to perfection when a tree falls over yet remains at least partly rooted. What were minor side branches become physically the new leaders, and get a new lease of life. They start growing as miniature trees, eventually producing a line of what appear to be individual trees. Incidentally, new or 'reassigned' branches tend to grow in the same shape as a seedling would; this programmed growth is often referred to as 're-iteration', repeating the same shape over and over again.

Biomechanics and gravity

Gravity plays a big part in determining the shape of individual trees. This can be understood by using the principles of 'biomechanics' fostered by a German physicist, Claus Mattheck. One of the simplest is the principle of minimum lever arms. This is less complicated than it sounds. As mentioned in the previous chapter, the longer a branch, the more likely it is to break (unless serious investment in wood takes place) because it is acting like a longer lever (in the same way that a longer spanner puts more turning force on a stiff nut). This acts to limit the ultimate length of a branch: a compromise between displaying leaves and keeping the lever to a minimum. There are, of course, ways of cheating: Chapter 4 decribes how banyans use pillar roots to support tremendously long branches, and if branches graft together (as in Figure 7.5) this support can allow them to become abnormally long.

These branches also act to bend the trunk. Trees on the whole grow to keep

Figure 7.5. Two oak (*Quercus robur*) branches grafted together. Just as roots pressing against each other will weld together, so will branches. This is often claimed to be impossible because the constant movement by wind prevents the tissues joining, but with enough pressure the union is possible, as demonstrated by the picture! Cheshire, England.

Figure 7.6. After the loss of the top part of the tree, a single branch acts like a lever bending the stem where the twisting force exerted (*M*) is the weight of the branch (*F*) multiplied by the length of the lever (*l*, branch base to the branch's centre of gravity). The tree reacts by bending the branch to reduce the length of the lever (the twisting force becomes zero), thus bringing the centre of gravity of the branch over the base of the trunk. This is analogous to the ease with which we can carry a bucket of water. From: Mattheck, C. and Kubler, H. (1995). *Wood — The Internal Optimization of Trees.* Springer, Berlin, Figure 18, Page 23.

the net effect of these levers to a minimum: in a normal tree the weight of the branches is balanced either side of the tree so that there is no net force acting to bend the tree. Figure 7.6 shows a tree that has lost its top leaving one branch sticking sideways. The power of the lever is calculated simply as the weight of the branch multiplied by the effective length of the lever. Very soon, the branch will begin to bend upright even when the tree is in the open and would get no more light by doing so. The advantage is obvious: it is to reduce the lever arm that is otherwise constantly putting a strain on the tree by pulling it sideways. The tree can detect the strain imposed by a lever arm and, other things being equal, will act to minimise the lever arm by bending branches by using the 'reaction wood' described in Chapter 3.

The same principle applies when whole trees lean. In Figure 7.7 the leaning tree will gradually bend to bring the centre of gravity above the base, to reduce the lever arm to a minimum. As the diagram shows, we do the same thing when carrying a heavy load on our backs and we bend forward to get the centre of

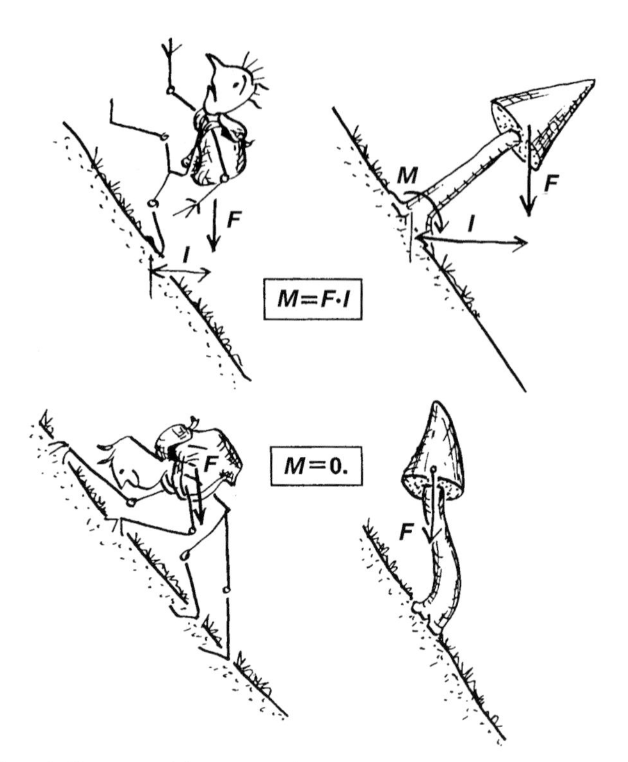

Figure 7.7. In a similar way to Figure 7.6, a leaning tree will straighten to reduce the twisting force exerted by gravity, in the same way that we bend forward when carrying a heavy pack to bring the centre of gravity over our feet. From: Mattheck, C. and Kubler, H. (1995). *Wood — The Internal Optimization of Trees.* Springer, Berlin, Figure 19, Page 23.

gravity above our feet. Otherwise we are constantly straining to stay upright and a little shove will have us over. In a tree this results in the characteristic 'J' or 'S' shape. At this point you may be thinking that you have seen plenty of trees that have been far from upright with the centre of gravity way off to one side. And you would be right. Sometimes, in the compromise determining tree shape, other things are more important; for a tree leaning out into the centre of a gap in the forest the extra light outweighs the strain of the lever. (Also, thick stems are less able to bend and a tree may be unable to bend to completely remove the lever arm.) You may be wondering how a tree with no brain works out these complicated solutions; all is revealed below in 'How does the tree control shape?'.

Buds, branches and tree shape

Standing on a cliff looking at trees shaped by strong winds leaves us in no doubt that growing conditions can alter the shape of a tree. But while individual trees may vary a lot depending on damage and growing conditions, the characteristic shape will shine through given half a chance, showing that it is genetically determined. How do trees manage to grow into such characteristic shapes? The answer lies in looking at the building blocks used. Animals, on the whole, have a fixed shape that simply gets bigger over the years. A baby and an adult human have all the same bits, they are just bigger in a grown-up. Plants on the other hand work by repeatedly adding together small modules, much like making a daisy-chain longer by adding more daisies rather than by making each daisy bigger. The basic module of a tree is most easily thought of as the leafy twig grown in one year.

A twig and its leaves

Starting at the end of a branch and following back, it is usually possible to see the bud scale scars forming a circle around the twig (and therefore sometimes called the girdle scar), which mark where last year's terminal bud was (Figure 7.8).

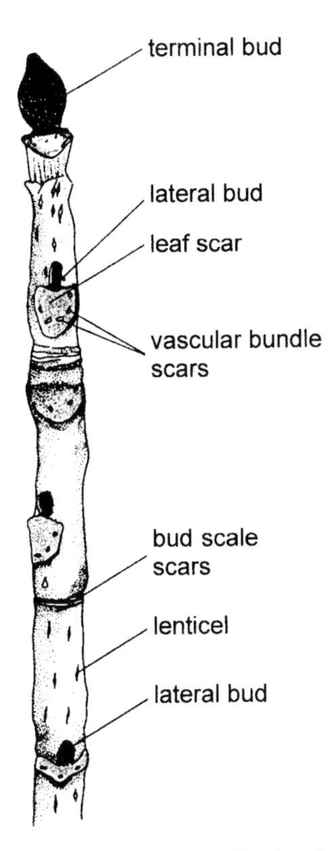

Figure 7.8. A branch in winter. The successive sets of bud scale scars mark how much of the branch grew each year. The leaf scars mark the positions of the previous year's leaves and the bud associated with each leaf is clearly to be seen. From: Nadakavukaren, M. and McCracken, D. (1985). *Botany: An Introduction to Plant Biology.* West Publishing, Saint Paul.

Thus, this end portion of the branch is what grew in the latest season and is the basic module of the tree. The young twig bears a number of leaves which, when they fall off, leave a leaf scar as shown in Figure 7.8. Nestling between the leaf and the twig is a bud (these are in the axils of the leaf and so are usually called axillary or lateral buds)[2]. These buds contain next year's growth, i.e. a new twig complete with leaves and buds[3]. It is worth pointing out here that new leaves can only be grown on new twigs; a bud never gives rise to just leaves although sometimes the twig may be inconspicuous—see the discussion on short shoots below. It is now possible to see how the tree gets bigger each year by each bud growing out into a new module, which bears buds, which grow out into further modules and so on. It is usually possible to trace your way back along a branch and perhaps the past 5–10 years' worth of growth, each year marked by the bud scale scars. This adding of new modules happens at the top of the tree and along each branch so the whole tree gets bigger. Bear in mind that the new twigs (modules) are the finger-like projections of the new coat of wood added over the whole tree so last year's twig will now be fatter and have two rings, the twig from the year before will be fatter still and have three rings, and so on, the branch getting thicker as you trace back to the trunk (see Figure 3.2).

Since the buds are in the axils of leaves, the arrangement of leaves (or the 'phyllotaxy') can play a large role in determining the basic branching pattern. Leaves and buds are basically arranged around a twig in a spiral but often this spiral is modified to produce leaves that are effectively alternate along a twig or, by reducing the length of twig between two or more leaves to a minimum, opposite or even whorled. Thus trees with alternate leaves (like oak and beech) tend to have alternate branches, and those with opposite buds such as ash and maple expand as a series of pairs of opposite branches, which can be traced back through the whole tree. Thus small-scale differences at the branch level influence the larger-scale shape of the crown.

A hopeless tangle

We have seen above that the tree is built up by continually adding new modules, just as a wall is built by adding new bricks, small in themselves but capable of adding up to a large imposing structure. But if every bud on a tree grew into a branch the canopy would soon become a hopeless tangle of dense branches. A 100 year old oak should have 99 orders of branching rather than the 5–6 that

[2] As always there are exceptions. In dawn redwood the bud is either beside or underneath the compound leaf. But since the 'leaf' is really a small twig with many leaves, perhaps it doesn't count (but is worth remembering for natural history quizzes!).

[3] Occasionally in fast-growing individuals of temperate trees such as birch, alder, cherry, and tulip tree (and more commonly in tropical trees) buds will grow out the same year as they are formed (referred to as sylleptic shoots as opposed to a shoot from a dormant bud, a proleptic shoot).

actually exist. Temperate trees rarely show more than 5–8 orders, and tropical trees 2–3 or at most 4 orders of branching (tropical trees generally have bigger leaves requiring a less fine network of branches to hold them). The potential tangle is prevented in three main ways: not using all the buds, shedding branches and altering the length of branches.

Dealing with too many buds

Sometimes the tree is helped by outside conditions. Spring frosts will, to our annoyance, readily kill buds when they are losing their cold hardiness. Abrasion of buds, leaves and twigs is also implicated as a cause of 'crown shyness' where the crowns of neighbouring trees (and branches on the same tree) rub together in the wind and do not intermesh. (White spruce—*Picea glauca*—growing under aspen in Canada can suffer in the same way from being whipped by swaying aspen branches.) But trees are not at the mercy of outside influences to solve the problem of too many buds. It has been shown in silver birch (*Betula pendula*), and is undoubtedly true of other trees, that fewer buds develop in parts of the crown that are already dense or where the crowns of different trees started meeting, presumably because of low light intensities. Moreover, more buds die in shaded areas and new branches are shorter. This helps explain why trees grown close together show a strong tendency to grow more on those sides that face away from neighbours, and another reason why trees show crown shyness.

Buds can also fail to develop as a normal part of the tree's growth. The spines of hawthorn and honey-locust, and the bundles of needles on pines, are all modified branches, which stop growth after a while and lose the ability to produce new buds. This simple solution of not producing buds is taken to the extreme in conifers. Conifers like spruces and firs with many small needle-like leaves do not produce a bud in the axil of every leaf[4]: if they did they would end up with an impossible number of buds and potential branches. Rather, they concentrate resources into relatively few buds at the end of each year's branches. As mentioned in the last chapter, some conifers, especially in the cypress family, do not produce any distinct buds at all!

Buds that *are* produced but are surplus to requirements can be deliberately aborted. In oaks there is a steady rain of aborted buds from the canopy throughout the growing season as up to 45–70% of buds are aborted. An alternative to getting rid of buds is to save them for later, keeping them for many years as an insurance policy in case of need (see Chapter 3). In oak the small buds at the base of a twig are usually kept in this way. Moreover, since bud scales are really modified leaves they themselves have axillary buds, which usually remain small

[4] Juvenile growth of pines, larches and cedars bears single needles that *do* have a bud in almost every axil. These buds produce the bunches of needles characteristic of mature growth. These needles in turn show the lack of buds in their axils as expected in conifers.

and largely unseen but quite capable of growth. The value of these stored buds is seen in pruning trials. The kermes oak of Mediterranean areas (*Quercus coccifera*) has been shown to produce new shoots from stored buds and survive for 4 years despite all new shoots being clipped off every 15 days. The lack of buds on many conifers shows in their difficulty in breaking from old wood if heavily pruned; this is especially noticeable in the cypress family, particularly Lawson's cypress (*Chamaecyparis lawsoniana*), which consequently makes a poor choice for hedging. Fortunately for landscapers, some conifers can form new buds in the apparently 'empty' leaf axils. Yew (*Taxus baccata*) and coastal redwood (*Sequoia sempervirens*) grow them and store them ready-made, while others, including the giant sequoia (*Sequoiadendron giganteum*) and white cedar (*Thuja occidentalis*) from N America, and the hiba (*Thujopsis dolabrata*) and Japanese red cedar (*Cryptomeria japonica*) from Japan, grow them when needed. Thus, like many hardwoods, they are capable of forming new branches when heavily pruned or damaged.

Loss of the terminal bud of a branch can have a large effect on the shape of a tree. If the tip lives for many years the branch will grow strongly in one direction (called monopodial growth) as seen in trees such as ashes. If it dies, the buds either side usually grow to replace it, giving a fork (sympodial growth), and if this happens many times the branch is made up of a series of short sections at angles to each other, as in mature horse chestnuts.

Shedding branches

A useful way of preventing too many branches clogging the tree is to get rid of them once they have fulfilled their purpose. This happens as the tree gets bigger and grows new shells of foliage which shade the inner and lower branches. If you look at a large tree the centre of the canopy is made up of large branches (the 'scaffold branches') leading to fine twigs only at the edge of the canopy. In the shaded centre (the 'dysphotic zone' in scientific parlance) the small branches that would once have occupied that space are long gone. Trees like the true cypresses (*Cupressus* spp.), coastal redwood, swamp cypress (*Taxodium distichum*) and the dawn redwood (*Metasequoia glyptostroboides*) regularly shed small twigs complete with leaves towards the end of summer. Other trees shed only those that prove unproductive. If a branch is not producing enough carbohydrate to cover its own running costs, i.e. it needs to be subsidised by other branches because, for example, it is being shaded, it will usually be got rid of. This prevents unproductive branches being a drain on the tree and removes the wind drag from useless branches.

Branches are shed for reasons other than lack of light. In dry parts of the world it is common for trees and shrubs to lose their smaller branches and save water. Small branches have the thinnest bark and greatest surface area, and are the source of most water loss once the leaves have been lost. The creosote bush of

USA deserts self-prunes in the face of extreme heat or drought, starting from the highest and most exposed twigs and working downwards to bigger and bigger branches; it's a desperate act because if it loses too much wood it dies. Shedding branches can, of course, also be useful for self-propagation. Most poplars and willows characteristic of waterways will readily drop branches, which root when washed up on muddy banks further downstream (see Chapter 8).

How are branches shed? In the simplest cases, dead branches rot and fall off, or healthy branches are snapped off by wind, snow and animals. Some willows have a brittle zone at the base of small branches that encourages breaking in the wind, seemingly for propagation. Other cases of 'natural pruning' are more startling; elms, and to a certain extent others including oaks, deodar (*Cedrus deodara*) and London plane (*Platanus* × *hispanica*), have a reputation for dropping large branches (up to half a metre in diameter) with no warning on calm, hot afternoons: hence the quote from Kipling 'Ellum [elm] she hateth mankind and waiteth'. Boy scouts should be warned not to pitch tents under elms! Such dramatic shedding appears to be due to a combination of internal water stress coupled with heat expansion affecting cracks and decayed wood.

Branch shedding in many trees is, however, a deliberate act. Branches are shed in the same way as leaves in autumn by the formation of a corky layer which leaves the wound sealed over with cork, which in turn is undergrown with wood the following year. Figure 7.9 shows the typical ball and socket appearance of branches shed in this way (officially called cladoptosis). In hardwoods, branches

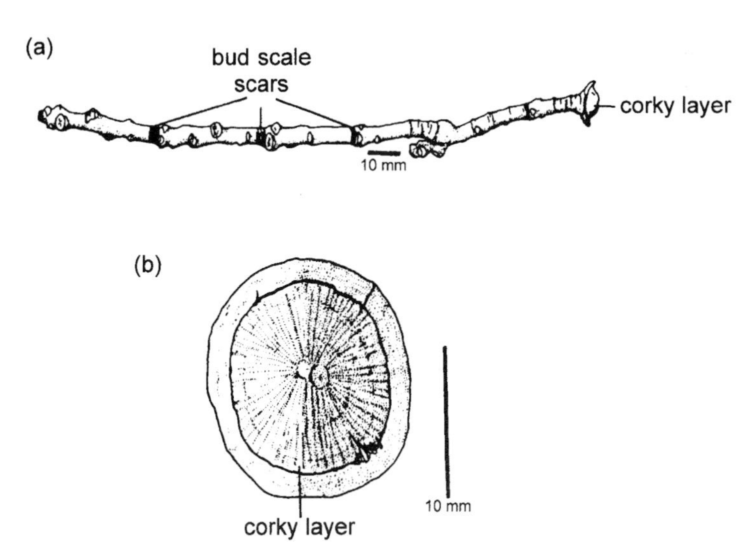

Figure 7.9. A branch shed from an oak (*Quercus petraea*) in the same way that a leaf is shed. Note the typical ball-shaped end (close-up in b) where the branch has been cut off by a corky layer. From: Bell, A.D. (1998). *An Illustrated Guide to Flowering Plant Morphology*. Oxford University Press, Oxford. Reprinted by permission of Oxford University Press.

up to a metre in length and several centimetres in diameter can be shed, normally after the leaves have fallen in the autumn (maples are unusual in casting branches mainly in spring and early summer). Oaks tend to shed small twigs up to the thickness of a pencil, beech may shed larger ones and birches dump whole branches of dead twigs. Pines shed their clusters of needles (really short branches) and members of the redwood family (such as the coastal redwood and giant sequoia, and the deciduous dawn redwood and swamp cypress) shed their small branchlets with leaves. Typically in hardwood trees, something around 10% of terminal branches are lost each year through a mixture of deliberate shedding and being broken off. Not surprisingly, shed branches can make up a third of the mass of forest floor litter.

Length of branches

Another way of reducing potential congestion is to make some branches smaller than others. As will be discussed further on, branches in the shade grow smaller than those in the sun. But trees can also regulate branch length from within. In many trees there is a clear distinction between 'long' and 'short' branches or shoots (Figure 7.10). This is true of apple, birch, beech, hornbeam, katsura, ginkgo, pines, larches and cedars (and perhaps also elm, poplar and lime, which have a more gradual transition between long and short shoots). The long shoots build the framework of the tree, making it bigger. The job of the short shoots (called spur shoots by horticulturalists) is to produce leaves, and commonly flowers, at more or less the same position every year. The short shoots grow in length each year just enough to produce closely packed leaves and the next set of buds: short shoots can therefore be told by the closely packed sets of bud scale scars from each year (see Figures 7.10, 7.13 and 7.14). The bundles of pine needles and the tufts of needles found in cedars and larches are also borne on short shoots. The cedar and larch short shoots continue to grow new needles for several years but in pines, the short shoot grows only about a quarter of a millimetre and stops, and then falls off intact once the needles die.

To maintain flexibility, any one shoot can swap from long to short or vice versa depending upon the internal control, light levels and damage (see below).

Flowering

Flowering can radically alter the shape of a tree because a growing point that ends in a flower dies when flowering and fruiting are over and cannot revert back to growing leaves. This is seen in magnolias, dogwoods, maples and horse chestnuts: the flowers are produced at the end of the shoot, which consequently does not produce a new bud. Thus, next year the two side buds (marked a in Figure 7.11) will grow out to resume branch growth, leaving a fork to mark where the flowers were (b in the figure), an example of the sympodial growth described above.

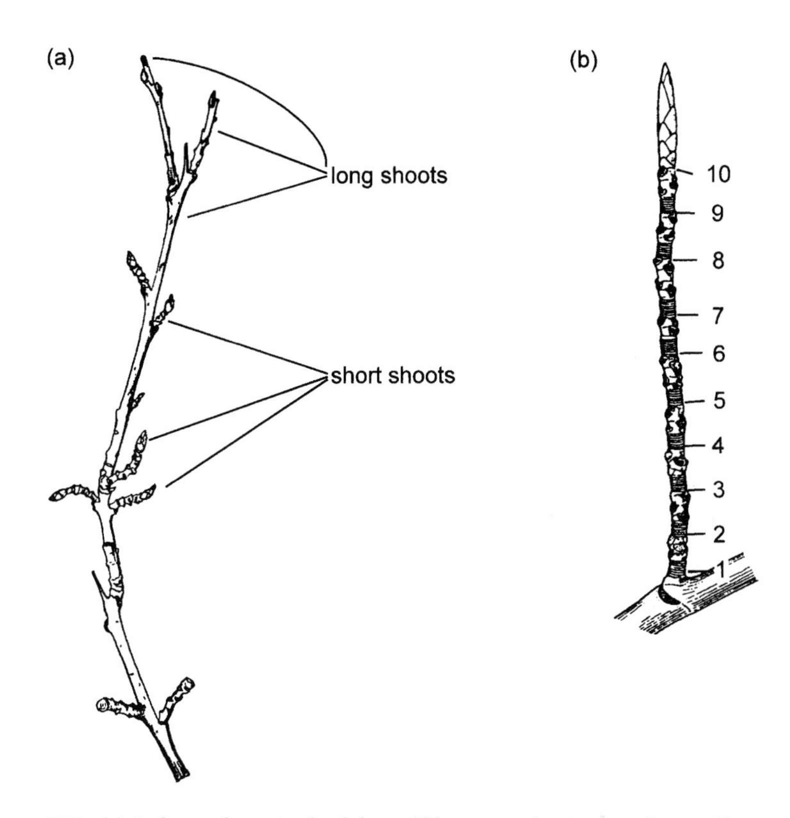

Figure 7.10. (a) A shoot of purging buckthorn (*Rhamnus catharticus*) made up of long shoots and short shoots; (b) a close-up of a ten-year-old beech (*Fagus sylvatica*) short shoot: annual growth can be identified by the crowded bud scale scars. From: (a) Büsgen, M. and Münch, E. (1929). *The Structure and Life of Forest Trees.* Chapman & Hall, London; (b) Troll, W. (1954). *Allgemeine Botanik.* Ferdinand Enke, Stuttgart.

The trees above are often described as fruiting on 'new wood', i.e. branches grown in the current year. In many trees, however, the flowers are tucked away further back on the branch, described as fruiting on 'second year' or 'old wood' (but bear in mind that the fruits, like leaves, are really hanging from a small twig grown this year). This is seen in, for example, elms and cherries. The terminal buds of each twig (marked a in Figures 7.12 and 7.13), and perhaps one or more further back on strongly growing shoots (b), produce just leaves, while the buds further back along the twig produce just flowers (c). These flowering points will, of course, not be able to produce new buds and so are used just once. Once the fruits are shed, that growing point is dead. Scars further down the branch bear testimony to previous flowering. The new leafy shoots grown this year will lay down buds to repeat the process next year.

Elms and cherries are fairly simple in that their buds produce either leaves or flowers. Other trees produce 'mixed' buds containing both flowers *and* leaves. This

Figure 7.11. Horse chestnut branch (*Aesculus hippocastanum*). The growing point dies once it flowers. Next year's growth will be from the two buds behind (a) leaving forks (b) to mark the site of flowers in previous years.

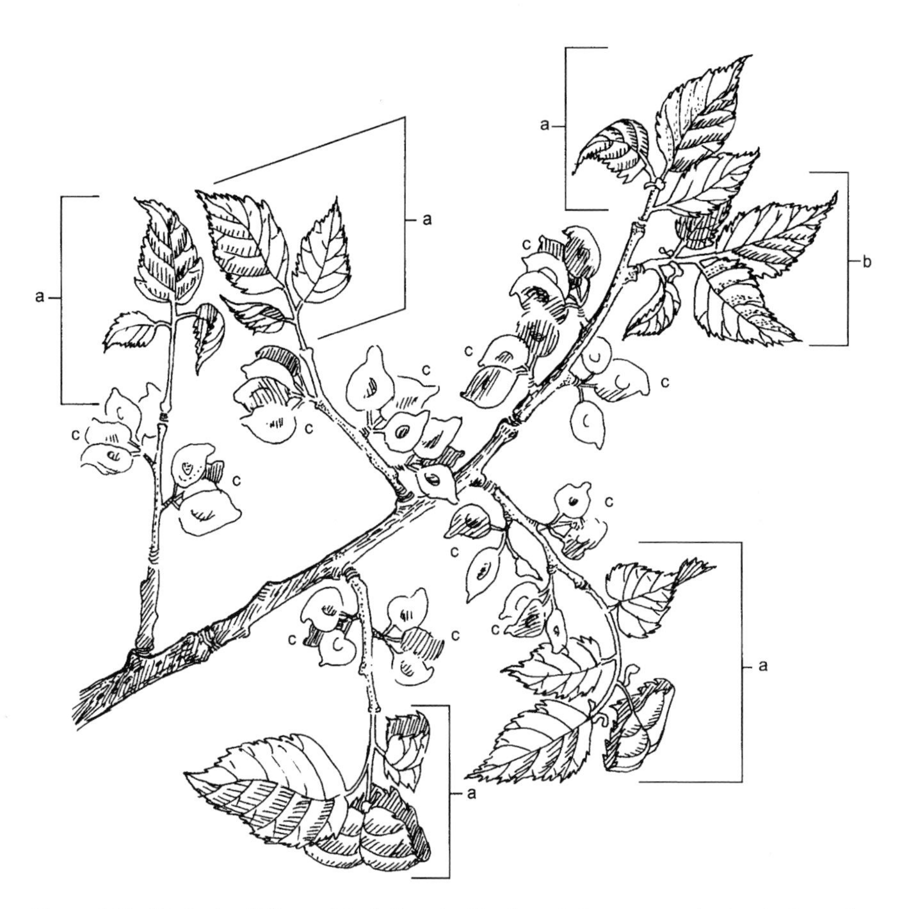

Figure 7.12. Wych elm (*Ulmus glabra*). Terminal buds, and the next one further back on the end shoot, have produced just leaves (marked a and b, respectively). Buds further back along the twig have produced just flowers (c).

is seen in the plum branch in Figure 7.14. Behind the buds producing just leafy shoots (a, b), the single 'mixed' bud at c has opened to grow both flowers (and hence plums) and leaves. Thus this bud has produced fruit *and* will be able to continue growing next year, unlike the flowering buds in elm and cherry. Further back on wood a year older, shoots show a mix of bud types. The terminal bud of the lowest twig shown is just a leaf bud (l), the other terminals are mixed buds (m), and the side buds further back on these shoots are just flower buds (f). The pear and blackcurrant also produce mixed buds but the flowers are at the end with a ring of leaves at the base so the growing point is still doomed to die. They get round this by using the bud associated with the lowest leaf for next year's growth. Other variations are possible, too numerous to mention here. It makes an interesting project to look at different branches and work out the strategy being used. The inevitable conclusion is that different patterns of producing

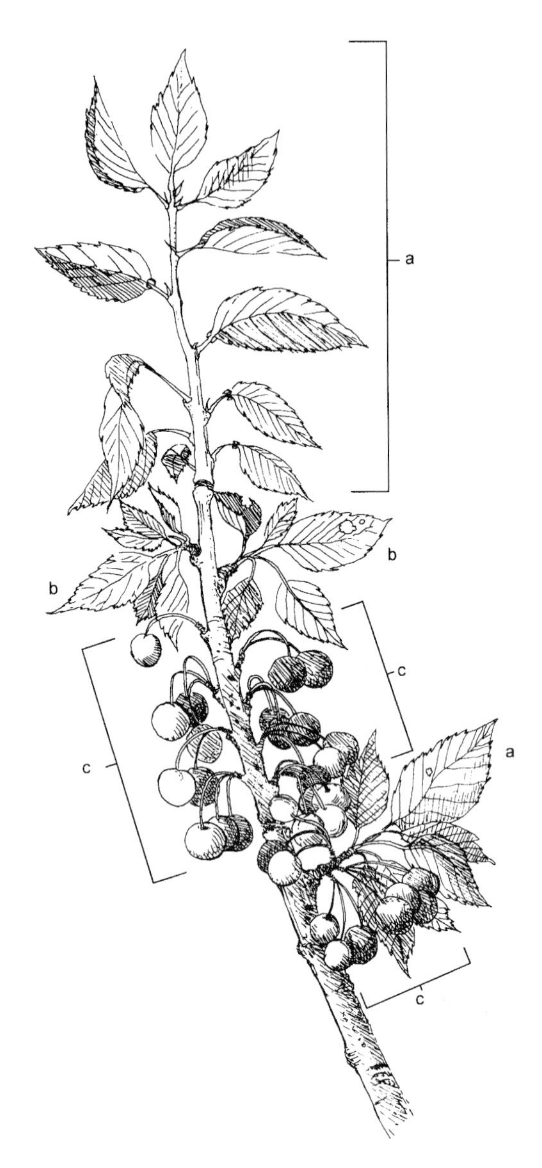

Figure 7.13. Cherry (*Prunus* variety) labelled as Figure 7.12.

flowers has a profound effect on the pattern of branches and hence the shape of a tree.

How does the tree control shape?

We have seen that the overall shape of a tree is governed by how buds and branches grow and die. How does the tree control this? The simple answer is that the ends

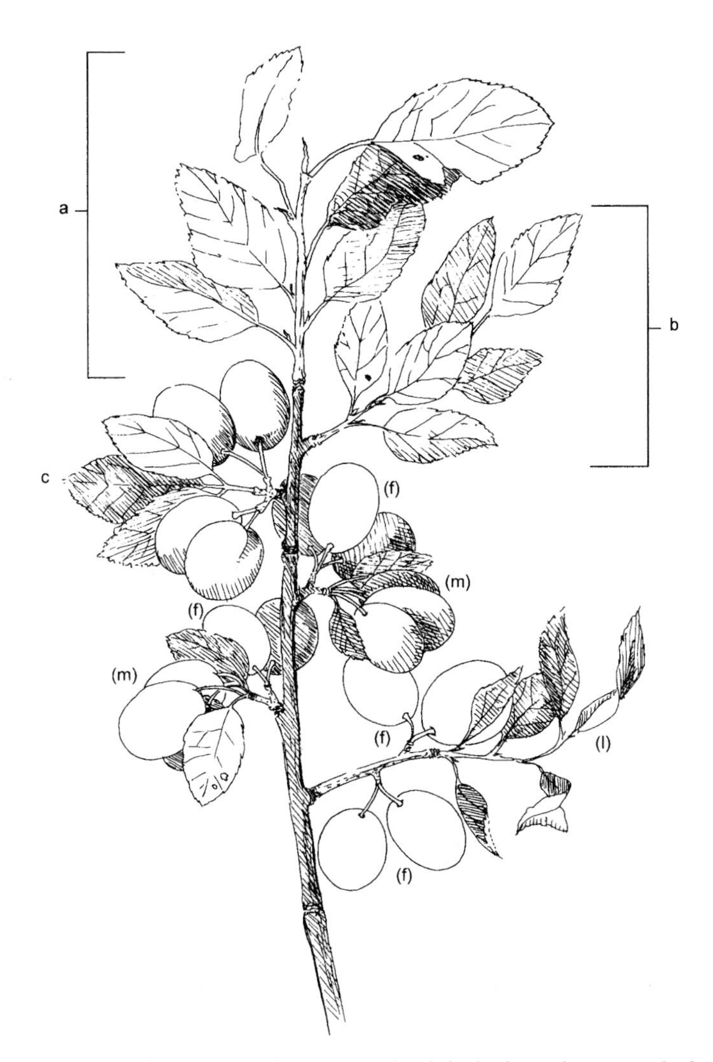

Figure 7.14. Victoria plum (*Prunus domestica*). Behind the buds producing just leafy shoots (a and b), the single 'mixed' bud at c has opened to grow both flowers (and hence plums) and leaves. Further back on wood a year older, shoots show a mix of bud types: flower buds (f), leaf (l) and mixed buds (m).

of the branches (apices in botanical parlance) tend to suppress what goes on further back along the branch. Over the whole tree this apical control can become quite complicated and is analogous to the power struggle within a kingdom. The topmost growing tip (the leader) is like the king having great power over the buds and growing tips immediately below. The strongly growing ends of side branches are like the princes of principalities and have great influence over the buds and branches near them but they themselves and their subjects are still under some control from higher up. In effect, every growing tip from the very top apex to the most minor branch end has some control, albeit diminishing in importance just like that from the king to a minor government official. The more minor the bud the more likely it is to be suppressed and the more likely that any resulting branch will be smaller and slower-growing (nothing we see in politics is new to nature!). The process is influenced from outside by the amount of light; the more light the greater the power of the bud and branch. Power does change with time. Like an ageing king, the leader(s) becomes less vigorous as the tree approaches it maximum height and loses some of the control, allowing the underling branches (princes) to exert more independence and growth. Complete removal of the seats of power by pruning (or accidental damage) means that one or more lesser branches (or new branches from suppressed buds) will usually bend upwards and compete to be the new leader. Clipping young Christmas trees releases dormant buds, making them bushier and more saleable.

Strictly speaking this method of control can be divided into 'apical dominance', which refers to suppression of buds from growing out, and 'apical control', which describes the control of subsequent growth from these buds. The reason this is mentioned is not to make your life more complicated but because it helps explain the two basic shapes of trees: the spire characteristic of the conifer and the broad oval of the hardwood. In rounded trees like the oak, apical *dominance* is very high so lateral buds are often suppressed. In future years, however, they are left far enough behind the apex that they are released and grow into branches. Because apical dominance *is* so high, these new shoots are very good at competing with the original tip and grow as fast or faster than the leader. This leads to the spire-shaped young tree disappearing into a mass of strong branches, giving the rounded shape (officially described as 'decurrent' which my Latin translates as 'running into', describing the way the branches merge). Conversely, in the tall spire of a conifer, apical dominance is weak, allowing the majority of buds to grow, producing the regular whorls of conifers. But since apical *control* is strong, the leader suppresses the growth of these new branches keeping them shorter and more horizontal, and the spire shape of all young trees is maintained for much of their life (an 'excurrent shape': the leader 'running out beyond'). The canopy is cone-shaped since the branches are longer lower down simply because they are older.

The mechanism by which buds regulate growth is not perfectly understood but the hormone auxin (produced by the apices) plays a fundamental role, possibly with other hormones such as cytokinins (initiators of bud growth in many species) and most likely aided by internal competition between growing points for minerals and sugars. These hormonal messages from the kings and princes are passed down the tree through the phloem. Horticulturalists have cashed in on this by producing chemical sprays that interrupt apical dominance and encourage bushy growth in ornamental plants (such as azaleas). Incidentally, a number of studies have shown that 'shoot inversion', bending the upper shoot over so that it points downwards, can release apical regulation. This undoubtedly explains the old country custom of snapping over the top branches of hazel to improve fruiting: hazelnuts are borne on short shoots, which will grow and fruit more prolifically once apical regulation is partly removed. Similarly, espalier-trained plants (where a vertical stem gives rise to several tiers of horizontally trained branches) is productive because tying a branch down closer to the horizontal makes it produce more fruit by weakening apical regulation (conversely raising a branch stimulates vegetative growth). This also explains how true weeping trees grow taller; as the leaders get longer and flop over, apical dominance is reduced and lateral buds are released, which produce new branches upwards, which flop over in turn. Thus weeping trees get taller in a 'stair-step' fashion (the exception are trees such as the deodar—*Cedrus deodara*—which straighten the pendulous tip during new growth).

As a finale in this section, is there any truth in the following ditty?

> *A woman, a dog and a walnut tree,*
> *The more you beat them, the better they be.*

Taking it as read that the first two are now rightly exempt, what about the walnut? In nineteenth century Europe, walnuts were knocked off the tree by using long poles. Branch ends were inevitably broken, which caused the production of more short shoots (by removal of apical regulation) and thus more flowers and fruit. Hence the custom of beating a barren tree to make it bear fruit. As mentioned in Chapter 3, beating the trunk may also help by bruising the phloem and so reducing the amount of sugars transported to the roots; these sugars are then available to be put into flowers and fruit.

Changes with age

The mature tree is not just a seedling writ large; the shape changes with age, so much so that it may be possible to age tropical rainforest trees purely by their shape. Young trees may be much less branched, especially if they have compound leaves, which are, in effect, cheap throw-away branches (e.g. horse chestnut) or if they self-prune (tulip trees, *Liriodendron tulipifera*, are easily identified in American forests because of the long straight unbranched trunk). In fact tropical rainforest trees can be up to 7 m tall before branching! As the number of branch ends increase, apical regulation is shared between more of them, leading to the rounded dome or flat top of maturity (as described above), and an increasing proportion of short shoots (more than 90% of growing shoots). How quickly the change in shape happens depends on species (early in horse chestnuts, late in poplars, never in many conifers) and soil (early on drier, poor soils).

Branch angle also changes with age. Over the short term branches have a set angle to which they will try to return by using reaction wood if bent by, for example, the weight of snow. But over the long term branches tend to sag as they get longer and heavier, helped by the compression of new wood in the branch crotch (in the same way that putting on more and more coats would force your arm to be more and more horizontal). Thus younger branches at the top of the tree are most upright and increasingly older branches down the tree are seen to be bent further outwards. The younger outer ends of old branches may object to being dragged down; old conifer branches that sag below the horizontal often have the young tip turned upwards. You might regard sagging branches in old trees as one of the disadvantages of age, like wrinkles and an expanding waistline, but it has been suggested that trees may benefit from these changes. The erect shape of a young tree will encourage rainfall hitting the tree to run down the stem, concentrating it on the developing roots at the base of the trunk. In an older tree, the sagged branches and weeping tips will encourage most water to be shed outwards and drip off the edge of the canopy onto the zone where their absorbing roots are concentrated. This is a gross oversimplification but does illustrate that many changes in shape do have a potential use.

Shape change with age can be even more radical. Perhaps the most classic is in ivy (*Hedera helix*), although it is admittedly a woody climber rather than a tree proper. As a juvenile it is a climber with heart-shaped leaves that clings on with aerial roots. Once the top reaches adequate light it changes to a mature, flowering phase with less deeply lobed leaves. It no longer produces aerial roots but holds its stems erect like any shrub and it flowers. This is such a definite change that if a cutting of the flowering shoots is rooted it carries on in the same fashion, making quite a decent upright bush! Trees also dramatically change shape as a matter of course by abrupt bending. Creeping willow (*Salix repens*), a plant of alpine Europe, starts life by growing vertically but soon bends at the base, thereafter growing prostrate. In some families of trees, and especially among conifers, cuttings taken from vertical and horizontal shoots (technically, orthotropic and plagiotropic shoots, respectively), follow the orientation of their original position. Thus a cutting from a horizontal branch will produce a weak tree with a tendency to be prostrate.

How are all the leaves exposed to light?

We come back to the problem of how a tree holds perhaps 100 000 leaves for each to catch the optimum amount of light. This would certainly pose a challenge for a solar-power engineer even without the added problems of changes caused by growing new leaves in different places and the constant movement by the wind! The first problem to solve is how leaves on a single twig avoid shading each other. In a vertical branch the leaves are carefully arranged so that they are not directly above each other (their positioning along a spiral can be described by the Fibonacci series, a mathematical progression that predicts the angles between successive leaves). Or, if they are likely to be directly above each other, as happens in a vertical branch of maple, the upper leaves are smaller with shorter stalks (Figure 7.15a). Preventing shading in a similar horizontal branch is not as easy but is solved by the upper leaf of a vertically oriented pair being smaller than its partner beneath (Figure 7.15b: this dissimilarity of leaves borne on the two sides is called anisophylly). This is even more pronounced in some tropical trees which on horizontal branches may have two rows of large leaves below and two of very small leaves above. The majority of trees produce a flat plane of leaves on a horizontal branch either by twisting of the leaf stalks or by alignment of the buds and leaves in this flat plane. Here the leaves avoid shading each other by neatly fitting together like pieces in a jigsaw puzzle (Figure 7.15c).

Looking at the whole canopy, some trees simply expand the method used for one twig to produce a single shell of leaves over the entire canopy, fitted together like a domed jigsaw puzzle. These 'monolayer' trees, such as beeches, hemlocks (*Tsuga* spp.) and sugar maple (*Acer saccharum*) are well adapted to shaded environments and are often found growing up beneath other trees in the comparative shade of a forest. On the other hand, trees growing in the open would waste a lot of useful light with just one layer of leaves: most leaves are working flat out in only about 20% of sunlight. To use the otherwise wasted light, layers of leaves are stacked one above the other to give 'multilayer' trees. It is not always easy to see the multilayers but the impression is one of a fairly open canopy which you can see into; for example, birches and poplars. These trees need a lot of light and so tend to do best in open areas. Fifty-four per cent of full sunlight appears to be the watershed; any darker and monolayer trees fare better, any lighter and multilayer trees grow better.

With more than one layer of leaves in a multilayer tree it is important that lower leaves are not sitting in the shadows of the one above otherwise they will not work very well. This is because the green light passing *through* a leaf has had the useful wavelengths of light largely removed and is little use to leaves below. The lower leaves need light that has passed *between* leaves. This could call for an impossibly high level of coordination of growth between different parts of the tree. Luckily, the solutions are simple.

1. Individual leaves can change their orientation to point towards where unfiltered light is coming from: possibly from the side.

2. As described in Chapter 2 under 'Sun and shade leaves', shaded leaves can use brief sunflecks of light penetrating a swaying canopy to good effect. Poplars and birches may gain from the fast-moving mosaic of sunflecks caused by the constant wobbling of their leaves on narrow leaf stalks.

3. If a leaf or branch is too shaded to be self-sufficient it will, as seen above, be shed.

4. If the layers are far enough apart the shadows of the upper layer disappear.

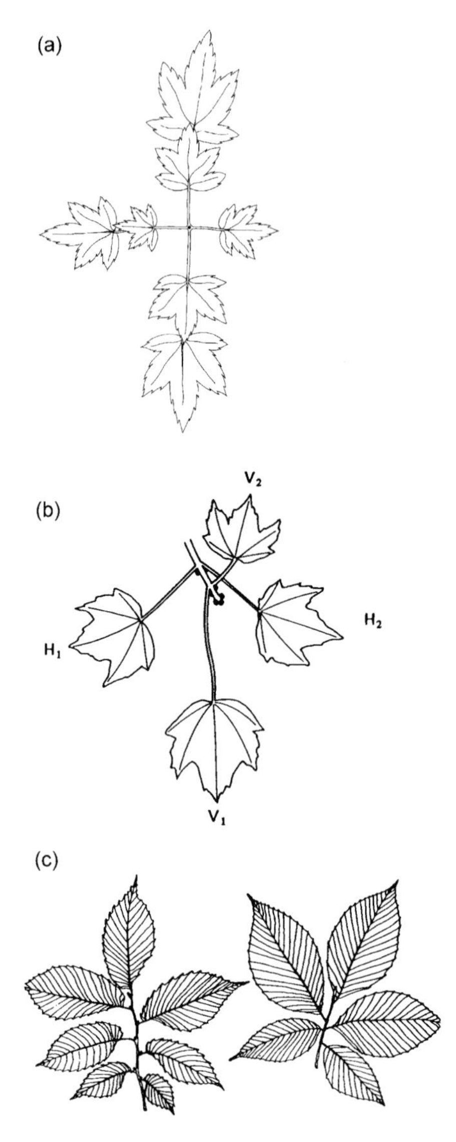

Figure 7.15. Arrangement of leaves on a branch to reduce self-shading. (a) In a vertical branch with leaves directly above each other (as in maples), the upper leaves are smaller with shorter stalks. (b) In a similar horizontal branch the problem is solved by the upper leaf of a vertically oriented pair (V_2) being smaller than its partner beneath (V_1). (c) In most trees a flat plane of leaves is produced on a horizontal branch where the leaves fit together like pieces in a jigsaw puzzle as in elm (*Ulmus* spp.) on the left and beech (*Fagus grandifolia*) on the right. From: (b) & (c) Zimmermann, M.H. and Brown, C.L. (1971). *Trees: Structure and Function.* Springer, Berlin, Figures III-27 & III-28a, Page 157.

This is because the sun is a disc rather than a point of light. You can try this with a light bulb and a small coin; hold the coin near a piece of paper and it casts a shadow, move it further away and although the light reaching the paper is dimmer, the shadow disappears. So providing the layers of leaves are far enough apart, the upper layer merely acts as a neutral density filter rather than casting a series of sun and shadow spots. Thus the leaves in the lower layer can be arranged independently of those in the layer above. The minimum distance between layers varies from 50–70 leaf diameters in sunny areas to just 1 leaf diameter in cloudy climates where the light comes from all over the sky (see Horn 1971 for further details).

Human influence

For centuries we have changed the shape of trees to suit ourselves. Sometimes this is for pleasure as in topiary where a range of supposedly aesthetic shapes are created by clipping such trees as privet (*Ligustrum* spp.), hiba (*Thujopsis dolabrata*) and yew. In other cases it is for food as in training branches (such as espalier work mentioned above) and pruning to encourage fruit growth and development (apples are bigger and redder if they and the leaves on the same spur (short shoot) are exposed to high light). Foresters have long controlled crown size and shape by adjusting the spacing of their crops. We have discovered chemicals which will stimulate or restrict growth: maleic hydrazide applied to new epicormic shoots will suppress their development in the following year; gibberellin biosynthesis-inhibiting growth regulators (e.g. paclobutrazol) have been used to reduce twig growth near power lines. We have also been busy producing trees of desirable shape through breeding and selective propagation (see Chapter 9). And we have changed tree shapes inadvertently. Perhaps the biggest influence has been our grazing animals, which are capable of keeping young trees trapped as shapeless bushes, if not killing them outright. Larger trees can have their lower growth pruned parallel to the ground, showing how high the local grazers can reach, whether they be rabbits, cattle or deer. Mechanical help is also to blame. Several roads near my home have beautiful lollipop-shaped hornbeams (*Carpinus betulus*) lining the verges, and most have a rectangular shape clipped out showing where high-sided vehicles have repeatedly pruned back the young growth! Mechanical help can also kill. Witness 'Sheffield blight'—damage done to bark by lawnmowers, etc., named in memory of Sheffield steel—which can do much to hasten the early demise of a tree.

Further reading

Berninger, F., Mencuccini, M., Nikinmaa, E., Grace, J. and Hari, P. (1995). Evaporative demand determines branchiness of Scots pine. *Oecologia*, **102**, 164–8.

Dewit, L. and Reid, D.M. (1992). Branch abscission in balsam polar (*Populus balsamifera*): characterization of the phenomenon and the influence of wind. *International Journal of Plant Science*, **153**, 556–64.

Fink, S. (1984). Some cases of delayed or induced development of axillary buds from persisting detached meristems in conifers. *American Journal of Botany*, **71**, 44–51.

Fisher, J.B. and Stevenson, J.W. (1981). Occurrence of reaction wood in branches of dicotyledons and its role in tree architecture. *Botanical Gazette*, **142**, 82–95.

Hallé, F., Oldeman, R.A.A. and Tomlinson, P.B. (1978). *Tropical Trees and Forests: an Architectural Analysis.* Springer, Berlin.

Horn, H.S. (1971). *The Adaptive Geometry of Trees.* Princetown University Press, Princetown, New Jersey.

Jones, M. and Harper, J.L. (1987a). The influence of neighbours on the growth of trees: The demography of buds in *Betula pendula. Proceedings of the Royal Society of London*, **B232**, 1–18.

Jones, M. and Harper, J.L. (1987b). The influence of neighbours on the growth of trees: The fate of buds on long and short shoots in *Betula pendula. Proceedings of the Royal Society of London*, **B232**, 19–33.

Maillette, L. (1982a). Structural dynamics of silver birch: I. The fates of buds. *Journal of Applied Ecology*, **19**, 203–18.

Maillette, L. (1982b). Structural dynamics of silver birch: II. A matrix model of the bud population. *Journal of Applied Ecology*, **19**, 219–38.

Neel, P.L. and Harris, R.W. (1971). Motion-induced inhibition of elongation and induction of dormancy in *Liquidambar. Science*, **173**, 58–9.

Putz, F.E., Parker, G.G. and Archibold, R.M. (1984). Mechanical abrasion and intercrown spacing. *American Midland Naturalist*, **112**, 24–8.

Simons, P. (1992). *The Action Plant.* Blackwell, Oxford.

Tsiouvaras, C.N. (1988). Long-term effects of clipping on production and vigor of Kermes oak (*Quercus coccifera*). *Forest Ecology and Management*, **24**, 159–66.

Watson, G. (1991). Attaining root:crown balance in landscape trees. *Journal of Arboriculture*, 17, 211–16.

Chapter 8: The next generation: new trees from old

In Chapter 5 we followed the processes of reproduction through to the arrival of the seed on the ground. Here we will look at germination and early survival of the seedling, and ways of producing new trees without resorting to seed.

The Seed

Seeds remind me of spaceships: they contain everything they need to colonise new worlds given favourable conditions and water once they arrive. The outside is covered by the seed coat (the testa), designed to protect the contents (see Figure 5.13). At the centre of the seed is the embryo, consisting of little more than a miniature root (the radicle) and shoot (the plumule). The rest of the seed is taken up with the food supply to keep the embryo alive before it germinates and to sustain early growth before photosynthesis takes over. This food is stored usually in the cotyledons (seed leaves), although some store it outside the cotyledons in the endosperm (which can be thought of as a short-lived half-brother of the embryo; all flowering plants have endosperm but in most it is used up quickly). In ash (*Fraxinus excelsior*), for example, the cotyledons are small, surrounded by endosperm, but in oak (*Quercus* spp.) the cotyledons are bloated and fill the seed with no remaining sign of the endosperm. The best example of endosperm is in the coconut (*Cocos nucifera*): part liquid (the milk) and part solid (the flesh). Wherever the food is stored, it is usually in the form of starch but oils are not uncommon especially in small wind-dispersed seeds. Oils contain more calories in a given mass and volume and so allow the seeds to travel light (having said that, large seeds can also contain oil: think of walnut oil in cooking and cocoa fat in chocolate). There is a price to pay: seeds with fats cost more to produce and are fairly short-lived (they go rancid.)

Seed dormancy

Most tropical species and a few temperate ones (notably elms, which produce their seeds in the spring; Chapter 5) have seeds that are ready to germinate as soon as they fall. But the seeds of most temperate trees are produced in the autumn and show some sort of dormancy to ensure they germinate in the spring rather than producing their delicate seedlings in the dying days of a warm autumn. In some, the seeds lie dormant for much longer and either germinate in a gentle dribble or await some signal to trigger mass germination.

The question then arises, what breaks the dormancy? Commonly, it is a period of winter chilling (but see Fire and heat in Chapter 9). Bear in mind that not all parts of the seed need chilling. In oaks and viburnums, for example, the root starts growing as soon as the seed falls in the autumn (protected from the harshest conditions by the soil) but the shoot only grows after chilling. This gives the seedlings a head start in the spring. As with bud opening (Chapter 6), a need for chilling can be circumvented; after a mild winter, germination is eventually induced by spring heat.

Dormancy can also be by the embryo being too immature to grow immediately after seed fall. The classic example is ash. Seeds fall in the autumn and after a winter chilling a few (less than 5%) may germinate but in the rest the embryo must mature over the following summer to germinate the following spring, around 18 months after the seed fell. Intuitively this would seem to put the seed at great risk of predation; there must be an over-riding reason.

A third cause of dormancy comes with the hard impermeable seed coat typical of the pea family and a number of others. The seed coat prevents the entry of water and germination cannot take place until the coat is ruptured in some way. This might be by fluctuating temperatures of hot summer days and cool nights, or the intense heat of a fire (e.g. gorse—*Ulex* spp.—and acacias), or rotting by fungi over time, or partial digestion when passing through an animal's gut. Yew (*Taxus baccata*) and juniper (*Juniperus* spp.) seeds, for example, germinate promptly once they have been through a bird's gut and then chilled but if dropped on the ground the thick wooden coat takes 1–2 years to rot enough for germination to be possible. In the same way the extinction of the dodo on Mauritius in 1681 has left the tree *Calvaria major* with no seedlings. It seems that the seeds, surrounded by a very thick fruit wall up to 15 mm thick, were eaten by dodos and retained in their gizzards for several days to help grind food until they had worn down enough to be either passed through the gut or regurgitated, ready to germinate. Temple (1977) tested this by feeding the fruits to turkeys; and it worked, producing the first seedlings for probably 300 years! Elephants, rhinos and bats have also been implicated in improving germination of tree seeds from typically 1–2% before eating to over 50% after, although in some cases this is due to the killing of insect predators that would otherwise eat the seed. But don't get the idea that hard seed coats are only found in exotic plants; even acorns will germinate more quickly and more completely if the seed coat is removed first.

Dormancy can be complicated. For example, people are often disappointed with the germination of holly (*Ilex aquifolium*); in exasperation they cut open a few seeds and finding no embryo conclude that the seeds are sterile. What is really preventing germination is a two-pronged dormancy. The hard coat prevents the entry of water (passage through a bird's gut helps). Once this is solved, the minute embryo develops and matures only after a prolonged warm period.

Gardeners have been imitating these natural processes for years. Seeds are given cold treatments ('stratified') by being mixed with damp soil and placed in a refrigerator; others are scarified by abrasion, nicked with a knife or soaked in acid to break the hard coat to induce these seeds to germinate.

A few woody species will not germinate until given light. These are usually small seeds that need a physical cue that they are not deeply buried or heavily shaded. And it's not just any old light they need. Light is detected in the seeds by the pigment phytochrome (see Chapter 6). Red light (wavelength around 660 nm) tends to break dormancy whereas far-red light (*c.* 760 nm) induces it. Light passing through leaves has a larger proportion of far-red light, so seeds needing light may be inhibited from germinating beneath other trees. This makes ecological sense: it is far better to sit and wait for a gap than face almost certain death as a small seedling in heavy shade. This switching effect of phytochrome may also explain why some trees have a light requirement for germination in the first place. Two ecologists, Cresswell and Grime (1981), working with herbaceous plants have shown that seeds surrounded by green fruits get a large dose of far-red light and a light requirement is induced; as the seed dries and becomes inactive this need is set. But where the fruit loses its green colour before the seed dries, seeds will have the phytochrome set in the active form and will not need light to germinate. Does this explain why most tree seeds have no light requirement because the thick fruit prevents far-red light reaching the seed?

In the same way that the need for cold can be circumvented, so also can the need for light. Birch will germinate in the dark if given heat.

Soil seed bank

If seeds falling to the ground remain dormant and do not germinate in the first spring, they will become incorporated into the reservoir of the soil seed bank. How long they remain viable depends on how long their food reserves last (all living things respire and burn up energy) and how long they survive the attention of predators and pathogens. Generally it is pioneer species that are most prominent in the seed bank: those species that invade open areas and are missing from the mature forest. These tend to have either small inconspicuous seeds (such as birches and tropical rainforest pioneers), easily overlooked by predators but which tend to be short-lived (2–5 years is probably average), or larger seeds with very good protection (such as cherries, gorse, brambles and raspberries) which may last 150–200 years. Densities can be high; in Europe, gorse (*Ulex europaeus*) and broom (*Cytisus scoparius*)—with hard seeds as described under seed dormancy above—have been found at densities of more than 30 000 and 50 000 seeds per square metre, respectively (although herbaceous species can approach half a million seeds per square metre).

Other seeding strategies

Remarkably few of the dominant woodland trees in Europe (such as oaks, beech, ash, etc.) have seed in the seed bank. Indeed, British acorns are readily killed by drying (termed 'recalcitrant') and therefore germinate quickly or not at all (as described above, the root grows out as soon as possible after seedfall and the shoot appears above ground the following the spring). Acorns come off the tree at around 45% moisture content and will cease to germinate if dried to 25% moisture (this can be reversed just a little by soaking in cold water for 48 hours).

Lack of soil storage is undoubtedly because these species have alternative strategies for optimising the chances of producing new trees. They may sprout from the base (elms) or roots (cherries), or be masting species (predator satiation is inconsistent with seed longevity; see Chapter 5). Also, many shade-tolerant trees, such as beech and the N American balsam fir (*Abies balsamifera*) and sugar maple (*Acer saccharum*), produce a mass of seedlings which can't grow to maturity in the shade but can persist for years as stunted seedlings. This 'seedling bank' gives a competitive edge over species starting from seed when a small gap opens up in the tree cover by a tree dying.

Fire and storage on the tree: serotiny

Not all trees drop their seeds when ripe, rather they store the seed up in the canopy (officially called 'serotiny'). A thousand or so species around the world do this including eucalypts, the she-oaks (*Casuarina* spp.) of Australia, the giant sequoia (*Sequoia sempervirens*) of N America, and many pines. Typically, the seeds are released by some sort of environmental trigger. This might be rainfall in desert shrubs and succulents but by far the most common trigger is fire. Just how and why serotiny works, and why it can be better than storing seeds in the soil, is discussed in detail in Chapter 9.

Seed size

The smallest seeds in the world belong to orchids. The Scottish orchid, creeping lady's-tresses (*Goodyera repens*), for example, has individual seeds weighing just 0.000 002 g, giving 50 million to the kilogram. Trees seeds are altogether larger and tend to be at the bigger end of the seed size range. As shown in Box 8.1, one of the smallest seeds of a woody plant is that of the heather of the British uplands at 33 million to the kilogram. At the other end of the scale there are fewer than a hundred horse chestnuts in a kilogram, culminating in the huge Seychelles double coconut, which weigh 18–27 kg *each* in a fruit up to 45 cm long. (Incidentally, the reputation of this palm for sea dispersal derives from the discovery of dead seed washed up on the Maldive Islands 2000 km (1200 miles)

Box 8.1. Average number of cleaned tree seed (i.e. without any of the fruit) per kilogram

Seychelles double coconut	*Lodoicea maldivica*	18–27 kg each!
Coconut	*Cocos nucifera*	4
Walnut	*Juglans regia*	85
Horse chestnut	*Aesculus hippocastanum*	90
Sweet chestnut	*Castanea sativa*	250
Common oak	*Quercus robur*	290
Pedunculate oak	*Q. petraea*	400
Hazel	*Corylus avellana*	1200
Sycamore	*Acer pseudoplatanus*	3300
Animal-dispersed pines		
Siberian stone pine	*Pinus sibirica*	4000
Swiss stone pine	*P. cembra*	4400
Whitebark pine	*P. albicaulis*	5700

Box 8.1. *(cont.)*

Beech	*Fagus sylvatica*	4500
Rowan	*Sorbus aucuparia*	5000
Ash	*Fraxinus excelsior*	14 000
Hornbeam	*Carpinus betulus*	24 000
Wind-dispersed pines		
Ponderosa pine	*Pinus ponderosa*	26,500
Weymouth pine/ Eastern white pine	*P. strobus*	58 400
Scots pine	*P. sylvestris*	200 000
Small-leafed lime	*Tilia cordata*	32 000
Ivy	*Hedera helix*	49 000
Wych elm	*Ulmus glabra*	73 000
Holly	*Ilex aquifolium*	125 000
Gorse	*Ulex* spp.	150 000

Box 8.1. *(cont.)*

European larch	*Larix decidua*	160 000
Norway spruce	*Picea abies*	195 000
Giant redwood	*Sequoiadendron giganteum*	200 000
Honeysuckle	*Lonicera periclymenum*	220 000
Coastal redwood	*Sequoia sempervirens*	260 000
Alder	*Alnus glutinosa*	770 000
Silver birch	*Betula pendula*	5 900 000
Aspen	*Populus tremula*	8 000 000
Dwarf birch	*Betula glandulosa*	8 450 000
Rhododendron	*Rhododendron ponticum*	*c.* 11 000 000
Heather	*Calluna vulgaris*	33 000 000

Based on data from:
Salisbury, E.J. (1942). *The Reproductive Capacity of Plants.* George Bell, London; Grime, J.P., Hodgson, J.G. and Hunt, R. (1988). *Comparative Plant Ecology.* Unwin Hyman, London; and Burns, R.M. and Honkala, B.H. (1990). *Silvics of North America.* Vol. I, *Conifers;* Vol. 2, *Hardwoods.* USDA Forest Service, Agricultural Handbook 654.

away, but fresh seeds sink in sea water! The coconut we buy to eat (*Cocos nucifera*) *does* float, however see Chapter 5.)

Germination

Once any dormancy has been broken and germination conditions are met (warmth, water and sometimes light), germination follows. As water is absorbed, the swelling seed splits the seed coat. The young root (the radicle) is first to emerge in search of a stable water supply, followed closely by the young shoot. As the shoot emerges above ground two things can happen to the cotyledons (seed leaves). In small seeds, as in pines and beech, the portion of shoot below the cotyledons usually expands rapidly, pulling the cotyledons, often still in their seed case, above ground where they will expand, turn green and start photosynthesising (Figure 8.1a,b). The young seedling then gets the best of both worlds; it uses the stored food and what the cotyledons can produce themselves. The cotyledons continue this dual function for normally 1–3 months until the first true leaves develop and the stored food is exhausted, where upon the cotyledons wither and fall. This type of 'epigeal germination' ('epi' means above) is found in many trees including maples, beech, ash, most conifers and small-seeded tropical species.

The alternative way of doing things is 'hypogeal germination' ('hypo', below). Here the shoot *above* the cotyledons grows, drawing the baby leaves above ground, leaving the cotyledons underground (as in oak, see Figure 8.1c). The cotyledons cannot, of course, photosynthesise and the plant has to wait until its first true leaves are functional before it grows any of its own food. This would seem to put hypogeal seeds at a great disadvantage in the race to the sun. The clue to resolving this paradox lies in seed size. Hypogeal germination is found in larger seeds: oak, walnut, horse chestnut, cherry, hazel, and the rubber tree (*Hevea brasiliensis*). It is risky to bring all that food above ground where it would be readily sought by herbivores, and anyway cotyledons are bulky objects to pull out of the ground (with possible damage to the delicate roots) and heavy to support in the air.

Despite their size, cotyledons carefully hidden below ground may not be as crucial as we might think to seedling survival. In Europe, acorns are primarily dispersed by jays (Chapter 5), which cache the acorns in the soil for winter fodder. Those that are not eaten will germinate and produce the next generation of oaks. But the jays can still find the seedlings in the spring. When they find one they give it a yank upwards, exposing the acorn containing the cotyledons, which are nipped off and eaten. Small seedlings may be uprooted but most survive without damage. Experiments have shown that seedling survival and growth is unaffected by such removal, even on nutrient-poor soil. It seems that once the first leaves are out, the seedling does not need the cotyledons. So why do oaks have such large acorns? The answer can be found below!

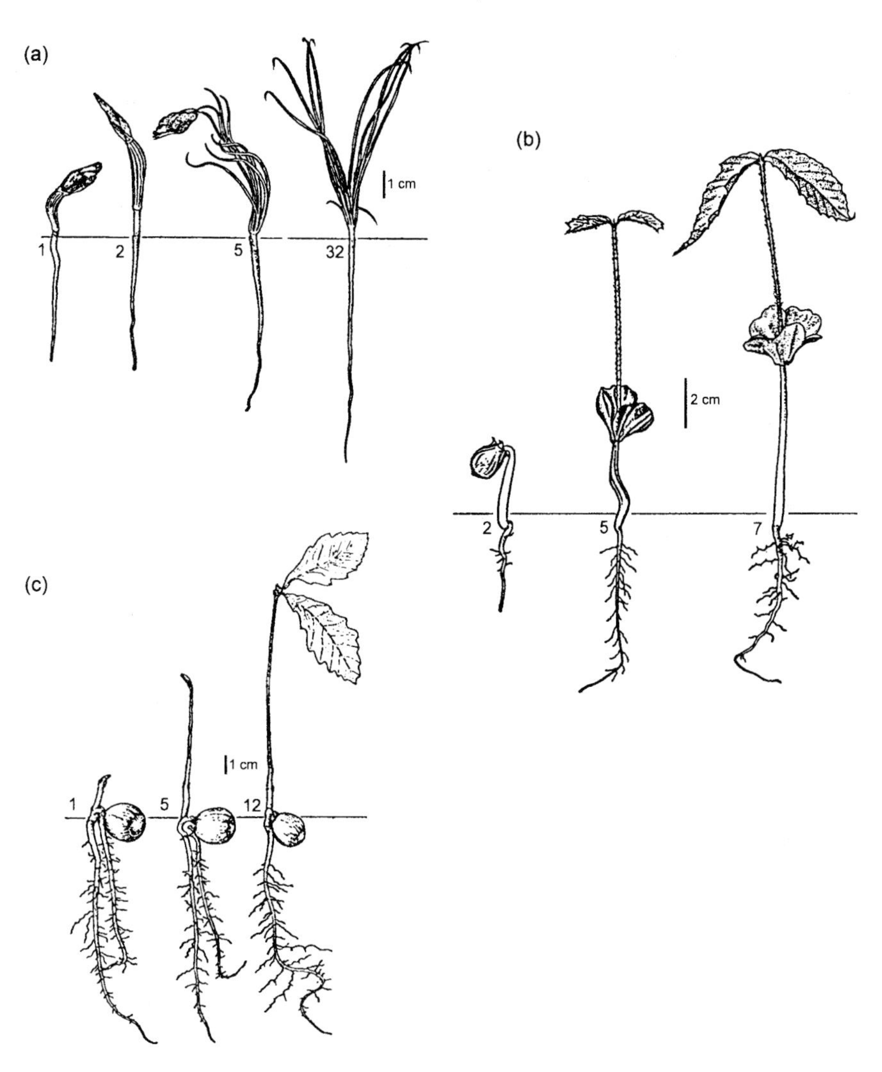

Figure 8.1. Seedlings of (a) pine (*Pinus palustris*) at 1, 2, 5 and 32 days after germination; (b) beech (*Fagus grandifolia*) at 2, 5 and 7 days; and (c) oak (*Quercus macrocarpa*) at 1, 5 and 12 days. Germination in pine and beech involves bringing the cotyledons above the ground surface ('epigeal') whereas in oak the cotyledons stay below ground ('hypogeal'). Pines have up to a dozen or even more cotyledons but the dicot hardwood plants (see Chapter 1 for an explanation) have two, and the advanced monocots have just one. Mistakes are not uncommon, though; sycamore will frequently have three or four cotyledons and in a handful of crab apple seeds I germinated last year several had an extra cotyledon. From: Anon. (1948). *Woody-Plant Seed Manual.* USDA Forest Service, Miscellaneous Publication No. 654.

The significance of seed size

The seed of the Mediterranean carob tree (*Ceratonia siliqua*) is so regular in size that it was used as the original measure of gold, the carat. Although most plants can't match this degree of regularity, nevertheless, seed mass is remarkably constant. Growing conditions will alter the number of seeds produced much more than their size. This indicates that size must play a pretty crucial role in plant establishment. First of all, though, seed size is not necessarily related to the eventual size of the tree: the worlds's biggest trees (coastal redwoods and giant redwoods—see Chapter 6) are well down the order of seed size (Box 8.1).

Generally, seed size is a delicate balance between on the one hand being small enough to maximise the number produced and (for wind-dispersed trees) maximise the distance they will spread, and on the other hand being large enough to give seedlings a good start. Not surprisingly, in pioneer species, which invade open areas (like birch and aspen), the emphasis is on small mobile seeds to ensure the greatest spread of seed. Conversely, trees that establish under dense shade (like beech and oak) do better with the bigger supplies of energy, nutrients and sometimes water found in large seeds. Larger seeds produce taller seedlings, which can more readily get above the shade of the woodland litter and herb layer (oak seedlings can grow 5–10 cm before the first leaves appear). Larger food reserves also enable a seedling to persist longer while waiting in the seedling bank for a light gap.

But coping with shade is the not the whole story. Large seeds are also of value in dealing with the deep, loose layers of undecomposed leaves found in woodland, which are notorious for rapid drying, leaving a shallowly rooted seedling high and dry. A large seed can grow a root rapidly, giving it a better chance of reaching the more constant water supply of the underlying soil before the litter dries to a crisp. Small-seeded pioneers tend to invade open areas bare of litter. And it is noticeable that where the climate is more constantly wet there are more smaller-seeded woodland species. Fast root growth also goes some way to explaining why maritime trees (like coconuts) are often large-seeded: they need to grow roots quickly down into water layers below that influenced by salt water. Indeed, early growth from the huge Seychelles double coconut (*Lodoicea maldivica*) puts the young embryo a half metre or so underground and up to 3 m sideways from the coconut—no wonder they're difficult to cultivate in pots! (This sinking of the seed's contents deeper into the ground has been termed 'cryptogeal germination' and is found in some African trees as a way of getting the seed out of the way of frequent fires.) The downside of large seeds is the expense of growing them. The Seychelles double coconut can bear only 4–11 fruits at a time and these take 10 years to develop.

We haven't finished with seed size yet because there are still several other factors that influence size. Large seeds are attractive bundles of food to herbivores and are not easily hidden and so must invest in costly defences. Eventually there comes a point where it is 'cheaper' to be smaller. This is nicely illustrated by a group of woody legumes in Central America that are plagued by the larvae of bruchid beetles, which live inside the seed. They divide neatly into those species with large seeds (an average of 3.0 g per seed) heavily protected by toxins, those with small seeds (average 0.26 g) which are readily attacked and rely on some being missed, and one species with very small seed (0.003 g) which is too small for a beetle larva to grow in and so go unmolested. Soil fertility also has potentially interesting evolutionary consequences on seed size. On poorer soils where the vegetation cover is slow to develop you would expect seeds of pioneer species to get smaller because they need fewer reserves to compete with neighbours and smaller seeds are more likely to spread wider and be more successful. For those invading dense forest on poor soils the opposite would be expected; heavier seeds, with a greater supply of nutrients, would tend to win the battle to dominate. This could explain the production of huge seeds typical on the impoverished soils of the tropics, so much so that the heavy fruits have to be borne on the trunk rather than thin branches (see Chapter 5).

This is complicated by the fact that the very largest seeds are probably an adaptation to attracting animal dispersers rather than anything to do with establishment. Here, we return to the jay. In Europe, oak is almost exclusively dependent on jays for dispersal and the jay prefers to hoard big acorns. So oak may have evolved big seed for dispersal rather than establishment (explaining why cotyledon removal has little effect on the seedling; see 'Germination' above). Similarly, many pines around the world have evolved to be spread by other corvid birds in the same manner; they have large wingless seeds (4–6 thousand seeds per kilogram; see Box 8.1) compared with 200 000 per kg in the wind-dispersed Scots pine.

The odds against success

An oak can produce more than five million acorns over its life, and a birch would easily reach hundreds of millions in its short life. For the most part, we are not over-run by these trees so the mortality rate must be high. As you would probably predict, the highest mortality is among the seeds and tender young seedlings, declining with age into adulthood. As a broad-brush figure the mortality of seed alone can often be around 95%. Of the 5% that germinate, another 95% may well die within the first year. These are very much general figures; the mortality of seeds and seedlings will obviously vary tremendously with year, site and chance events (like landslides and volcanic eruptions) but figures of the order of one in a thousand acorns becoming a seedling, and one in a million acorns becoming a mature oak are probably worth considering when thinking of survival rates. Why such tremendous wastage, especially when we can plant a few seeds in a plant pot and most of them will grow? The answer lies with a fundamental problem of plants: they can't move. Seeds have to be planted by using animals or physical agents like wind as intermediaries, which is inevitably a hit-and-miss process. How would our gardens look if we planted them by throwing handfuls of seed out of a window? Add in the ravages of disease and herbivores and the need for such a large number of seeds to get just a few trees becomes a pragmatic reality.

Periodic establishment of new seedlings is by no means uncommon. This may be because seedlings can only establish when a tree dies, creating an opening, in dense woodland. Or it may need some sort of environmental disturbance such as fire, avalanches, wind storms, earthquakes and volcanic eruptions to play a part. The worrying lack of seedlings of some southern beeches (*Nothofagus* spp.) in Chile, which live for four or five centuries, was found to be because they establish after periodic devastating earthquakes. We must make sure that we look at seedling establishment on the time scale of the trees. For an oak living more than 500 years, a few decades of no success is a drop in the ocean. At the same time, we must not be complacent about lack of new seedlings if we humans have had a part to play. The lack of oaks in past decades has been claimed to be due to the persecution of jays (thankfully largely a thing of the past) and removal of large native herbivores, which would have created good seedbeds. Certainly, the 'recent' lack of seedlings in oak and beech woods has been blamed by some on the cessation of woodland grazing by cattle and pannage of pigs.

In temperate forests dominated by just a few evenly spread species, the survival of seedlings is often fairly random (and thus independent of seedling density; see Shabel and Peart 1994). In others, and especially tropical forests, survival can be linked to how many other seedlings are near by (density-dependent survival; see Condit *et al.* 1994). In some cases it is those packed together that survive best and in others it is those furthest apart! The former—survival in clumps—works on the same principal as survival of seeds in masting: predator satiation (Chapter 5). Large numbers of seedlings in a small area will overwhelm the ability of local herbivores to eat them all and some will therefore survive. Since most seeds do not travel very far, the seedlings end up clumped around the parent tree. The opposite—seedlings surviving when away from their fellows—is found where herbivores roam large areas searching for specific types of seedling and will feed voraciously when a clump is found. Thus it is the odd isolated seedlings at the outer edge of the dispersal area that are most likely to be missed and survive to grow. This 'escape hypothesis' may go some way to explain why individuals of most tropical trees are usually well spaced through a forest and not clumped as in many temperate forests.

New trees without seed: vegetative propagation

There are several ways that new trees can be produced from pieces of an existing tree. Suckers (which are shoots from roots) are common on a number of temperate trees such as elms, cherries and plums, false acacia (*Robinia pseudo-*

acacia), sweet gum (*Liquidambar styraciflua*), aspens and poplars. Suckering in the last two happens fairly readily and allows them to invade sites not suitable for establishment from seed. In this way great stands of aspen and poplar have developed on the floodplains of N America, covering many hectares each (Chapter 6) with thousands of trees, all clones of (i.e. genetically identical to) the original tree(s). Some of the other trees listed are less prone to suckering unless damaged. A cherry tree outside my garden was heavily pruned five years ago and had some of the roots disturbed by digging. The tree has been throwing up suckers ever since within a radius of 10 m of the trunk.

Branches touching the ground may root and produce new upright stems by 'layering'. Like suckering, this is rare in the tropics and most common in high latitudes and alpine areas where damp peaty soil is common. Conifers seem to be best at it: spruce, juniper, fir, yew, hemlock, false cypresses in the north and podocarps in the south (but only rarely in pines). Broadleaved trees can also layer naturally (e.g. forsythia and the lowest shoots of coppiced hornbeam) but this is on the whole rare (although not impossible to do artificially: layering is a handy means of propagation in those trees hard to root from cuttings).

New trees can also be produced by twigs breaking from or being shed by the tree (see Chapter 7), sticking in the ground and growing as natural cuttings. This obviously works best in wet places and may explain why poplars and willows down any one river tend to be either all male or female rather than the mixture you would expect from seed. Poplars and willows certainly drop branches that are in peak health, full of nutrients and with plenty of buds: perfect propagation material. The only trouble is that few studies have found new trees growing from broken branches in the wild!

As we have seen before, most hardwoods and a handful of conifers can regrow from the stump if the canopy is damaged, removing apical regulation. Old or newly formed buds (epicormic and adventitious buds, respectively; Chapter 3) grow into new stems using the stored food and roots of the old tree. However, since this just replaces the previous tree it is debatable whether you would call this propagation.

New trees can be produced artificially by micropropagation (or tissue culture). Basically, cells taken from a tree can be grown in the laboratory into new individual trees, each genetically identical to the original tree (cloning). In this way, good trees can be reproduced without the gamble of genetic variation inherent in seed production. Such techniques have been used to propagate timber trees such as cherry, and fruit trees such as dates.

Producing new *types* of tree

Cuttings (especially from young growth) can be rooted to create new trees. Although this cannot produce new species, it can create new growth forms (or cultivated varieties, shortened to cultivars). This relies on the way a cutting 'remembers' its way of growing. Thus cuttings taken from the top of a tree produce fast, upright growth while side branches once rooted will often carry on growing sideways, giving a prostrate shrub or at least a slow-growing shrubby specimen. By taking cuttings from the tops of mature hollies or false acacias where there are few spines (temperate herbivores can't reach that far and nature doesn't waste effort) we can create thornless varieties.

New varieties are also produced by watching for genetic mutations or 'sports'. Very occasionally a seed will produce an unusual specimen (a seedling sport) owing to a mutation in its genetic code. Copper beech arises this way and a number have been found growing wild in continental Europe (though most seedling sports will only survive in the cosseted nursery bed). If you want one of these lovely beeches, however, it is easier to take a cutting than having to plant thousands of seeds in the faint hope of striking it lucky (even seed from a copper beech produces mostly green or just faintly coloured seedlings).

More commonly the odd unusual branch can be found on a tree (a bud or branch sport) which if rooted as a cutting will produce a whole tree of the sport type. Pink grapefruits, seedless grapes and navel oranges all come from single such branches. If you consider that a large tree may have between 10 000 and 100 000 growing points (buds), even a relatively low mutation rate of around 1 in 10 000 produces a large number of sports. Most mutations are detrimental and the branch quickly dies. But occasionally one will be benign enough to merely lead to an interesting variation, which we find worth propagating. Similar mutations can happen over and over again, resulting in confusingly similar varieties hitting the market. Mutations can produce dwarfness, odd shapes, colour variation (think of the large number of variegated and golden conifers) and juvenile fixation where juvenile foliage (mostly on conifers; see Chapter 2) is kept right through to maturity (that's how we get the 'Squarrosa' and to a lesser extent the 'Plumosa' varieties of the sawara cypress, *Chamaecyparis pisifera*). Note that trees propagated from bud mutations tend to be unstable and revert back to produce vigorous foliage of the original tree that should be cut out to prevent it taking over.

Occasionally a cutting from a 'witch's broom' in a hardwood will produce dwarf trees. The causes of witches' brooms (the ball-like dense tangle of small twigs looking almost like a round bird's nest) are largely unknown, but see Chapter 9.

Hybridisation (reproduction between usually closely related species) is another way of getting new types of tree. Often this happens naturally but we are not beyond encouraging it. Perhaps one of the most famous deliberately produced hybrids is the hybrid larch (*Larix* × *eurolepis*[1]), a hybrid between the European

[1] A hybrid between two species in the same genus is denoted by a cross before the new species name (e.g. *Larix* × *eurolepis*). A hybrid between two species in separate genera is shown by a cross in front of the new genus name (e.g. × *Cupressocyparis leylandii*). In the same way the + sign is conventionally used to indicate a plant that is a graft hybrid.

and Japanese larches beloved by foresters because of its hybrid vigour, growing more vigorously than either parent at least while it is young. But probably the most famous hybrid, if the large number cluttering up urban Britain is any indication, is the Leyland cypress (× *Cupressocyparis leylandii*), a cross between two species in different genera, the Nootka cypress (*Chamaecyparis nootkatensis*) and the Monterey cypress (*Cupressus macrocarpa*). The original Leyland cypress hybrid occurred at Leighton Park in Wales in 1888 where both parents happened to be growing near each other (in their natural range along the western seaboard of N America they don't come within 300 miles of each other). Six plants were raised from seed collected from a Nootka Cypress by C.J. Leyland. In 1911 the reverse cross happened: seeds collected from a Monterey cypress produced 2 seedlings. Leyland cypress shows hybrid vigour and the tallest is over 35 m and growing strongly. At this rate it has been suggested that they have the potential to become the tallest trees in the world and we'll all be pale troglodytes! Unlike animals, hybrids in plants are often fully fertile. In tree families where hybridisation is common (the birch, pine, rose and willow families) this has led to back-crosses with the parents to produce a wide range of intermediates (a hybrid swarm) creating a nightmare for field botanists and students facing an identification exam!

Grafting can also be used to create new types of trees, or at least new sizes. In grafting, the top of one tree (the scion) is grafted onto the bottom of another (the stock). This allows a weak plant to benefit from the strong roots of another, or a vigorous tree (such as an apple) to be kept small by growing on 'dwarfing rootstock'. A big problem with grafts is the need to cut back vigorous shoots from the stock, which might swamp a delicate scion. One also needs to ensure that the stock and scion will grow in diameter at the same rate, otherwise strange-looking trees result (Figure 8.2).

Whether you consider grafting to be a way of producing new types of tree is arguable but grafts can give rise to something far more interesting. The scion is joined to the stock only at the point of grafting. But occasionally a bud arising from the union point grows into a strange mixture of the two plants, which literally has the core of one plant and a thin skin of the other: a graft-hybrid or graft chimaera (from the Greek mythological fire-breathing monster with the head of a lion, body of a goat and tail of a dragon). Probably the best known woody graft-hybrid is the laburnum–Spanish broom hybrid, + *Laburnocytisus adamii*. The original specimen (from which all subsequent trees have come as cuttings) arose in France in 1825 in the garden of a nurseryman called Adam who grafted the Spanish broom (*Cytisus purpureus*) onto a laburnum (*Laburnum anagyroides*). The tree is a laburnum at its core, a fact which is reflected in its overall shape, but the skin, just one cell thick in this case, is Spanish broom and the leaves and twigs are broom with a hint of laburnum in the shape. The flowers can be broom purple or laburnum yellow or a mixture, sometimes divided down the middle into half yellow–half purple! In general the

Figure 8.2. Problems with grafted ash trees (*Fraxinus* spp.). In the foreground the rootstock is growing in diameter faster than the grafted-on scion, and in the background the reverse is happening. Photographed at the Royal Botanic Gardens, Kew.

flowers of graft hybrids are either sterile or produce seeds of the parent at the core (e.g. laburnum). These trees are usually stable but some buds may develop into branches of the core species. Thus in the cut-leaf or fern-leaf beech (*Fagus sylvatica* 'Asplenifolia'), where a core of ordinary beech has a skin of the cut-leaf form, branches with ordinary beech or intermediate shapes are not uncommon.

To make life more interesting, there are other types of chimaeras. Variegated plants are chimaeras formed naturally within one plant by mutation in the growing point. In most plants the growing point or meristem is composed of discrete layers of cells like the pages in a book each of which arises from one initial cell. A mutation in an initial cell will be spread through the resulting layer. Bud mutations that result in sandwiches of colourless and green layers produce the white-edged variegated leaves commonly seen in pelargoniums and a range of trees from maples to hollies.

Further reading

Bossema, I. (1979). Jays and oaks: an eco-ethological study of a symbiosis. *Behaviour*, **70**, 1–117.

Condit, R., Hubbell, S.P. and Foster, R.B. (1994). Density dependence in two understory tree species in a neotropical forest. *Ecology*, **75**, 671–80.

Cresswell, E.G. and Grime, J.P. (1981). Induction of a light requirement during seed development and its ecological consequences. *Nature*, **291**, 583–5.

Dewit, L. and Reid, D.M. (1992). Branch abscission in balsam poplar (*Populus balsamifera*): characterization of the phenomenon and the influence of wind. *International Journal of Plant Science*, **153**, 556–64.

Foster, S.A. (1986). On the adaptive value of large seeds for tropical moist forest trees: a review and synthesis. *Botanical Review*, **52**, 260–99.

Galloway, G. and Worrall, J. (1979). Cladoptosis: a reproductive strategy in black cottonwood? *Canadian Journal of Forest Research*, **9**, 122–5.

Gosling, P.G. (1989). The effect of drying *Quercus robur* acorns to different moisture contents, following by storage, either with or without imbibition. *Forestry*, **62**, 41–50.

Grime, J.P. and Jeffrey, D.W. (1965). Seedling establishment in vertical gradients of sunlight. *Journal of Ecology*, **53**, 621–42.

Harper, J.L., Lovell, P.H. and Moore, K.G. (1970). The shapes and sizes of seeds. *Annual Review of Ecology and Systematics*, **1**, 327–56.

Judd, T.S. (1994). Do small myrtaceous seed-capsules display specialized insulating characteristics which protect seed during fire? *Annals of Botany*, **73**, 33–8.

Kelly, C.K. (1995). Seed size in tropical trees: a comparative study of factors affecting seed size in Peruvian angiosperms. *Oecologia*, **102**, 377–88.

Krasny, M.E., Vogt, K.A. and Zasada, J.C. (1988). Establishment of four Salicaceae species on river bars in interior Alaska. *Holarctic Ecology*, **11**, 210–19.

Lamont, B.B., LeMaitre, D.C., Cowling, R.M. and Enright, N.J. (1991). Canopy seed storage in woody plants. *Botanical Review*, **57**, 277–317.

Leck, M.A., Parker, V.T. and Simpson, R.L. (1989). *Ecology of Soil Seed Banks*. Academic Press, New York.

Rees, M. (1993). Trade-offs among dispersal strategies in British plants. *Nature*, **366**, 150–2.

Shabel, A.B. and Peart, D.R. (1994). Effects of competition, herbivory and substrate disturbance on growth and size structure in pin cherry (*Prunus pensylvanica* L.) seedlings. *Oecologia*, **98**, 150–8.

Sonesson, L.K. (1994). Growth and survival after cotyledon removal in Quercus robur seedlings, grown in different natural soil types. *Oikos*, **69**, 65–70.

Temple, S.A. (1977). Plant–animal mutualism: coevolution with dodo leads to near extinction of plant. *Science*, **197**, 885–6.

Tilney-Bassett, R.A.E. (1986). *Plant Chimeras*. Edward Arnold, London.

Thomas, P.A. and Wein, R.W. (1985). Delayed emergence of four conifer species on postfire seedbeds. *Canadian Journal of Forest Research*, **15**, 727–9.

Vázquez-Yanes, C. and Orozco-Segovia, A. (1992). Effects of litter from a tropical rainforest on tree seed germination and establishment under controlled conditions. *Tree Physiology*, 11, 391–400.

Chapter 9: Health, damage and death: living in a hostile world

It's a tough world. Trees face a constant battle in competing for light, water and minerals with surrounding plants. As if that were not enough, they also have to fend off the attention of living things, which view trees as good to eat and places to live. Insects chew away on all parts of a tree and are quite capable of completely defoliating it. Larger leaf-eating animals (which are usually on the ground since a belly full of compost heap is a heavy thing to carry around; leaf eating monkeys are an exception) chew away at the lower parts of the tree, although giraffes can reach up around 5.5 m. Whole armies of animals that can climb and fly will feed on the more nutritious flowers, fruits and the sugar-filled inner bark (see Chapter 3). The grey squirrel, introduced to Britain from N America in the 1880s, is a prime example. This rodent does extensive damage to hardwoods by stripping bark in spring to get at the sweet sap. It seems that dense stands of self-sown hardwoods have little sap and are largely immune (which may be why it does not cause problems in its native home) but well-tended planted trees have thin bark and a high sap content and are mercilessly attacked. So big is the problem that ash, lime and wild cherry may become more common in Britain because of their relatively low palatability to squirrels at the expense of palatable beech and sycamore.

Other animals are not adverse to making their homes from or in trees. Many birds, including various eagles, have been seen tearing off sizeable branches for nest making. Others live completely off the tree. Gall wasp larvae stimulate their host plant to grow galls on whichever part each species specialises in (leaves, flowers, fruits and stems). The nutritious tissues swell up around the grub and give it a home and food supply. Witches' brooms, the mass of densely branched small twigs that resemble a besom lodged in the canopy, are grown in the same way, induced by a range of different organisms: sometimes by fungi (e.g. *Taphrina betulina* in birches), mites, or even in N America by dwarf mistletoes (such as the *Arceuthobium* species). To this list of parasites you can add a number that normally live off the tree's roots and are only seen when they flower, such as the broomrapes (*Orabanche* spp.) and toothwort (*Lathraea squamaria*) of Europe. European mistletoe (*Viscum album*), in contrast, is only partly parasitic; it just takes water from the tree's plumbing and grows its own sugars by using its green leaves.

Epiphytic plants such as lichens and ivy merely use the tree as a place to grow up near the sun. In theory they take nothing from the tree. However, there is some evidence that tropical epiphytes, including a range of bromeliads and orchids, may intercept nutrients washed from the tree's leaves and branches by rain which would otherwise eventually reach the tree's roots. This 'nutritional piracy' may be significant to tree health in the tropics and explains 'canopy roots'—Chapter 4. A word about ivy (*Hedera helix*). It has long been debated whether ivy kills trees. I stand in the camp that says ivy poses no problem for a healthy tree. The problems come with old weak trees where the sail area created by the evergreen ivy can make the tree vulnerable to windthrow in the winter when winds tend to be strongest. Likewise, it is rare for the ivy to swamp the leaves of a healthy tree. The light intensity inside the canopy—ash is an exception—is low enough to prevent the flowering ivy shoots from growing (see Chapter 7); and it is these flowering shoots growing upwards like bushes that are most likely to compete with new growth.

As if all this was not enough, diseases also take their toll. Fungal diseases, especially of the roots and stems, are particularly important in tree health (more on this later). Bacteria and possibly viruses also play a role. For example, 'wetwood' is a bacterial rot creating pockets of moist rot with plenty of methane.

Defences

Being firmly rooted in the ground, a tree cannot move to escape harsh conditions or the unwanted attentions of animals and diseases (seeds are the only parts with an option for movement). And there are no two ways about it, because trees are so long-lived they face a tremendous number of problems over their lives. They have therefore developed an impressive array of defences to protect themselves where they stand. The living skin of a tree is normally no older than two or three decades, and the leaves, flowers and fruits are usually even shorter-lived. Consequently, the defences of these parts are similar to those in non-woody plants. The real specialisation of defence comes in maintaining the woody skeleton which may persist for centuries or even millennia. We'll consider these defences in turn.

First-line defences: stopping damage

Physical defences: spines, thorns and prickles

Large herbivores are capable of eating copious quantities of leaves along with the odd twig. Indeed the low nutritional value of leaves *requires* large volumes to be eaten. The chief defences against these large herbivores are spines, thorns and prickles. Spines and thorns (which botanically are the same thing) are modifications of a leaf, part of a leaf, or a whole stem (Figure 9.1). In *Berberis* species (as in cacti) it is the leaf that is turned into a spine; this leaves the branch with no green leaves so new ones are produced from the bud in the leaf axil growing out this year rather than next as is normal. Holly (*Ilex aquifolium*) could be

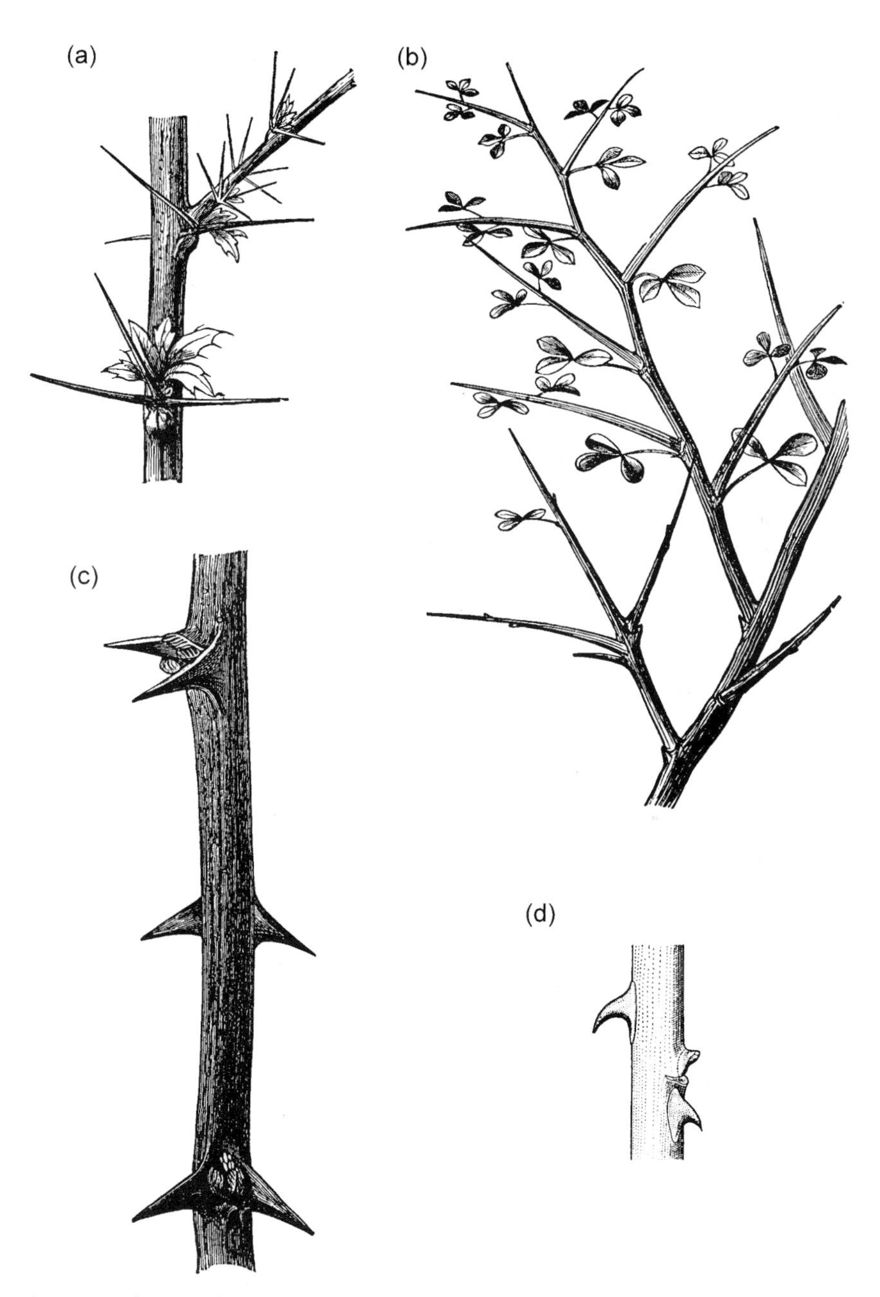

Figure 9.1. Thorns and prickles. (a) Barberry (*Berberis* spp.): thorns made from modified leaves. (b) A spiny broom (*Cytisus spinosus*): thorns made from modified branches. (c) False acacia (*Robinia pseudoacacia*): thorns made from the two stipules at the leaf base, which remain on the twig when the leaf falls. Prickles (d) as found on the stems of a rose (*Rosa* spp.). From: (a)–(c) Oliver, F.W. (1902). *The Natural History of Plants.* Vol. 1: *Biology and Configuration of Plants.* Gresham, London. (d) Troll, W. (1959). *Allgemeine Botanik.* Ferdinand Enke, Stuttgart.

regarded as a half-way house with just the margin growing spines. In false acacia (*Robinia pseudoacacia*) it is just the stipules (outgrowths of the leaf base) that are modified into persistent thorns. Hawthorns (*Crataegus* spp.) and firethorns (*Pyracantha* spp.) have thorns *above* the leaf, showing that the thorn is a modified branch, grown out from the bud. Growth next year can still take place because these trees cunningly grow extra buds beside the thorns. Again there are half-way houses; some brooms (*Cytisus* spp.), blackthorn (*Prunus spinosa*) and other *Prunus* species produce thorns at the ends of normal branches.

The prickles of roses (*Rosa* spp.) and blackberries (*Rubus* spp.) are quite different from thorns (Figure 9.1), being merely outgrowths of the bark or skin (and so with no vascular tissues) which can be easily broken off. While being of use for defence they also help these scrambling plants to hook onto whatever they're climbing over.

Spines and thorns are expensive things to produce and will only be grown where they are needed. Since most leaf-eaters stand on the ground to eat, simply because they are too heavy to climb or fly, it is perhaps not surprising that trees like holly produce leaf spines mostly on the lower 2 m of the canopy. But why do thorns appear in some trees but not others? Peter Grubb of Cambridge University proposes a solution to what appears to be a haphazard distribution. First of all, physical defences are found in habitats where nutritious growth is scarce. This explains why thorny plants are found in deserts and heathlands (e.g. gorse). But it also explains why thorns are found on plants that invade gaps in forests, e.g. hawthorns, apples, honey-locust (*Gleditsia triacanthos*), and false acacia (*Robinia pseudoacacia*): food is scarce because other vegetation is out of the reach of large herbivores. The same principle applies to European holly, which, being an evergreen in a background of deciduous trees, provided scarce fodder in the winter. (Indeed around the southern Pennines in England, shepherds would cut the upper branches of holly (no prickles!) as fodder, particularly in spring when other foods were not available.) Finally the spines common around the growing points of palms and tree ferns are well worth the investment because they have only one growing point which, if damaged, means death to the plant (see Chapter 3).

Other physical defences

Trees have not been slow in evolving other physical defences. Hairs help deter insects attracted to young vulnerable growth. In some cases this is a physical barrier (like us having to chew through a pillow to get food), in others the hairs contain chemical deterrents (as in the nettle-like hairs of the Australian rainforest stinging trees, *Dendrocnide* spp.).

Other physical defences can be quite subtle. For example, because of the arrangement of their mouths, caterpillars have to eat a leaf from the edge; so holly has evolved a thickened margin to prevent the caterpillars getting a start.

So why haven't more trees evolved a similar mechanism? Probably because the cost is only worthwhile in long-lived evergreen leaves and these are usually protected by chemicals instead. Physical defence can be yet more subtle. One example of many is the subterfuge indulged in by passion flower vines (*Passiflora* spp.), which are eaten by the caterpillars of heliconid butterflies. The butterflies rarely lay their yellow eggs where there are already plenty (this would be pointless competition for food amongst the caterpillars) so the vine grows imitation yellow eggs on its leaves!

Ants and other beasties

Around the world, there are many examples of tropical trees that use ants as their main defence. The ants run around the tree preventing birds and animals from nesting, dissuading herbivores, cutting away epiphytes and lianas, and in some cases killing anything growing within a 10 m radius of their tree. These ants usually have vicious stings as is rapidly found out if you lean against their tree! In return the tree provides food and often a home as well. For example, the 'ant acacias' of the New and Old World have swollen thorns in which ants hollow out nests. Nectar is delivered from leaf stalks ('extrafloral nectaries', i.e. nectaries not in flowers) and protein from bright orange bodies produced at the tips of the leaflets. Ant acacia leaves are less bitter than in other acacias: presumably the cost of looking after ants is cheaper than producing internal deterrent chemicals.

Employing guards can work on a much smaller scale. It is common to find little tufts of hair at the junctions where veins join together on the underside of leaves. A study on the European evergreen shrub *Viburnum tinus* found 10 species of mite—mostly predators and microbivores (i.e. eating microbes)—living in the crevices (or 'domatia') between these hairs. In other words, the plant is providing homes for mites that help keep other harmful organisms at bay.

Chemical defences

Woody plants produce a great variety of chemical compounds to provide protection against other plants, diseases and herbivores big and small. These include alkaloids, terpenes, phenolics (e.g. caffeine, morphine, tannins and resins), steroids and cyanide producers (cyanogenic glycosides). These chemicals can be found in just about any part of the plant (for example, the waxy surface of apple leaves contains toxins to repel certain aphids). They can also be emitted into the air in considerable quantities which explains why pine woods smell so nice and the eucalypt-dominated bushlands of Australia have a blue haze. These chemicals seep out through the bark or the holes in a leaf (the stomata) or ooze out of special glands, often on teeth around the leaf edge (as found in pines, alder, willows, hornbeam, maples, ashes, elms and viburnums) usually when the leaf is young but sometimes (as in crack willow) even late in life. Are these emissions of expensive chemicals just inadvertent leaks or do they have a purpose? Certainly in oaks it allows the trees to communicate with each other: just how is revealed below.

Chemical defences work in several ways. Some are highly toxic and will kill attackers in small doses (referred to as 'qualitative' defences since they work by being what they are rather than how much is there). These include the alkaloids in rhododendron (honey from its flowers is poisonous to humans but is so bitter it's uneatable), and cyanide (hydrocyanic acid) produced, for example, by crushing leaves of cherry laurel (*Prunus laurocerasus*) and consequently used to effect in an insect-killing bottle. Other chemicals work by building up in the herbivore (called 'quantitative' defences since they become more effective the more there is). Classic examples are phenolic resins in creosote bushes and tannins found in a number of plants. Tannin is a 'protein precipitant' which makes food less digestible so herbivores end up starving and stunted. These often have the effect of causing the herbivore to move elsewhere. Incidentally, it is possible that the endangered British red squirrel is declining not through competition with the introduced N American grey squirrel but because of acorns. Acorns contain digestion inhibitors that greys can disarm but reds can't. Red squirrels do well in conifer plantations feeding on the more nutritious pine seeds and where there are no oaks to give the greys a competitive edge.

Plants are not immune to the effects of chemical defences. Spanish moss (*Tillandsia usneoides*) that festoons trees in southern USA and South America (and is not a moss but a sophisticated relative of the pineapple) never grows on pines, presumably because of the resins. Perhaps the best known example of anti-plant chemical defences was first reported by Pliny the Elder in the first century AD, who wrote that 'The shadow of walnut trees is poison to all plants within its compass'. Walnuts (and all trees in the genus *Juglans*) contain the chemical juglone, which, seeping from the roots, reduces the germination of competitors, stunts their growth and even kills nearby plants, resulting in open-canopied walnuts having very little growing under them. Tomatoes, apples, rhododendrons and roses are very susceptible but many grasses, vegetables and Virginia creeper (*Parthenocissus* spp.) will happily live under walnuts.

Defensive chemicals are costly to produce and store (many are toxic to the plant producing them). Oaks may put up to 15% of their energy production into chemical defences (which explains why stressed trees are most prone to attack: they have less energy available to produce defences). Thus, although there may be large pools of defences swilling around (preformed) where attacks are common and ferocious, it makes evolutionary sense in lesser situations to produce the chemicals in earnest only when they are needed (induced). Oaks (and a number of other trees) will tick along with low concentrations of tannins in their leaves but once part of the canopy is attacked, the tree will produce tannins in large quantity. More than that, the tannins released into the air are detected by surrounding oaks and they in turn will produce more tannins in preparation for the onslaught. In a similar way, willows in Alaska browsed by snowshoe hares produce shoots with lower nutritional concentration and higher levels of lignin and deterrent phenols, thus rendering them less palatable. These 'induced' defences may have an effect for some time; birches can remain unpalatable for up to three years after being browsed.

Expensive defences are saved not just *when* they are not needed but also *where*. The N American aspen (*Populus tremuloides*) has been shown to produce fewer tannins in trees growing on fertile soils: it is cheaper to replace lost material than to defend it. Similarly, more is invested in defences where herbivores and diseases are more common. This is certainly true of tropical forests, where 7–8% of annual growth is consumed by herbivores compared with 3.5–4% in temperate forests. This explains why so many interesting chemicals come from tropical trees, to which we are not immune! Some can be particularly unpleasant, doing to us what they are designed to do to the animals that venture to eat bits of them. Strychnine, from the bark of various *Strychnos* lianas (primarily *Strychnos nux-vomica*, native to Burma and India) causes twitching and muscle spasms that can cause the head to be forced back to the buttocks, breaking the spine. Only death from exhaustion releases the victim from indescribable agony. Curare from the bark of a number of lianas (including *Strychnos toxifera* and *Chondodendrum tomentosum*) paralyses muscles (a healthy herbivore defence!), which can cause death by asphyxiation and heart failure but which fortunately is also a boon in aiding surgeons to operate on relaxed patients. The vicious arsenal is not restricted to the tropics. Seeds of various members of the rose family (apples, plums, peach, apricot, almond) contain doses of cyanide as hydrocyanic acid. Leaves of the California bay (*Umbellularia californica*) contain cyanic acid and when handled give off an unpleasant smell, causing headaches and even unconsciousness!

Other chemicals are less toxic to use humans and we even find them quite pleasurable. Where would we be without the defensive chemical caffeine found in coffee and tea plants, or the chemicals in the bark of the cinnamon tree? Indeed some have saved lives. In 1535 Jacques Cartier was stranded in the ice-bound Canadian St Lawrence River near modern-day Montreal. Twenty-five men were already lost to scurvy when passing Iroquois Indians taught them to boil the bark and foliage of what we now call white cedar (*Thuja occidentalis*) to make a tea. The high vitamin C content of the tea apparently worked miracles. Cartier named the tree 'arbor vitae', tree of life. Similar stories using other woody plants as antiabscorbants are many, from Captain Cook using spruce beer, to Captain Vancouver using Winter's bark (*Drimys winteri*) from South America. (Incidentally, unpasteurised cider has a high vitamin C content, a good excuse for imbibing a little of the pale fluid if considering a sea voyage!) Vitamin C seems to do the same job for plants as for animals: it is an antioxidant keeping plants safe from their own aerobic metabolism and a range of pollutants. I can't pass without mentioning that the ancient Greeks used finely ground pine cones mixed with herbs as a cure for haemorrhoids.

The evolutionary battle

No defence is impenetrable. Over evolutionary time herbivores find ways around a defence and the plant counter attacks with new defences. And so the battle goes on. Yew is poisonous to cattle and horses and yet deer will happily eat it. Red and fallow deer eat holly despite the prickles. Deer and rabbits don't eat sycamore bark but squirrels love it. Willow leaf beetles not only survive the salicin (raw aspirin) from their hosts, they use it themselves to produce a defensive secretion. Caterpillars can eat through a whole tree of young oak leaves despite the tannins (the caterpillars have an alkaline gut, which reduces the tannins' effectiveness). The tree responds by shedding the damaged leaves to prevent disease entry and the waste of further nutrients on substandard leaves.

It is puzzling how a tree ever survives in the battle of attacks and counter-attacks. A tree living for centuries with a genetically fixed array of defences will be faced with insects that produce thousands of generations during this time with thousands of opportunities for producing new ways of getting around the tree's defences. Part of the solution may well lie in the sheer size of a tree. A large oak or pine carries something between 10 000 and 100 000 buds (meristems). A naturally occurring genetic mutation in one of these growing points will result in a whole branch that is genetically distinct from the rest of the tree. Even though rates of mutation are low and most mutations are likely to be deleterious (and will be shed as dead twigs and branches), genetically distinct branches should accumulate over the years, turning the tree into a giant genetic mosaic. This may explain why some branches tend to lose their leaves earlier in autumn. It is *possible* that some of these distinct branches will be better resistant to the evolving insect attack than others, and so prosper: natural selection operating within a tree. As these resistant branches grow most vigorously they will make up a larger proportion of the canopy. They will also be the branches to produce most seed, which will be genetically best adapted to cope with the current set of pests. In this way, long-lived trees can go some way towards decoupling their rate of evolution from their generation time and give them a chance against rapidly multiplying pests.

Defending the woody skeleton

At first sight it may seem that wood does not need much defending: it's pretty inedible stuff. The large quantities of cellulose (40–55%), hemicellulose (25–40%) and lignin (18–35%) are all tough carbohydrates that are quite hard to decompose. Moreover, wood is incredibly poor in protein and hence nitrogen (typically 0.03–0.1% nitrogen by mass compared with the 1–5% found in green foliage). Just how poor wood is as a food is illustrated by the goat moth, whose caterpillars burrow through wood and take up to four years to grow to maturity, and yet they can mature in just a few months if fed on a good rich diet.

Despite the starvation diet that wood offers, there are many insects, fungi and bacteria that are capable of living off wood.

Keeping things out: resins, gums and latex

These fluids have a primary role in rapidly sealing over wounds (whether created by insects or by physical accidents) and in deterring animals from forcing their way in. Any animal rash enough to burrow is physically swamped and trapped, and may be overcome by chemical toxicity (though bark beetles are seen swimming through resin apparently unharmed, if a little hindered).

Resins If you have leant against an old pine or handled a cone you will be well aware of the ability of conifers to produce copious quantities of resin. The typical 'pine' smell of conifer foliage comes from the resin. Yews (*Taxus* spp.) do not contain resin and consequently do not smell.

In most conifers, including pines, Douglas fir (*Pseudotsuga menziesii*), larches and spruces, the resin is contained in ducts that run through the bark and wood, tapering off into the roots and needles. Others such as the hemlocks, true cedars and true firs, have resin restricted more or less to the bark, although, like other conifers, they are capable of producing 'traumatic resin canals' in the wood after injury or infection. In the true firs (*Abies*) the resin is contained in raised blisters. Cells along the ducts or blisters secrete resin, creating a slight positive pressure, so if the tree is damaged the resin oozes out. Once in the air, the lighter oils evaporate, leaving a solidified scab of resin over the wound.

Different species of tree produce different resins. At the beginning of the American civil war, Union forces in the Californian foothills were cut off from their normal supply of turpentine so they distilled it from the resin of the ponderosa pine (*Pinus ponderosa*). But at higher altitudes there grows the very similar Jeffrey pine (*Pinus jeffreyi*) whose resin contains high levels of heptane, a highly inflammable liquid found in raw petroleum. Firing up a primitive turpentine still full of Jeffrey pine resin was like building a fire under a petrol tank!

Trees other than conifers produce resins. Members of the family of Burseraceae contain resins particularly in the bark. This includes frankincense (*Boswellia carteri*), used in incense and chewed by Arabs as a breath freshener, and myrrh (*Commiphora* spp.) used in incense and perfumes.

Gums Fulfilling a similar function, a wide range of woody plants produce gums. The family of Anacardiaceae is notable for gum-producing trees including the varnish tree (*Rhus verniciflua*), a native of China, whose gum is used as the basis of lacquer. Gums are also found oozing from wounds in a variety of temperate trees such as those in the genus *Prunus* (the cherries, plums, etc.). These gums are carried in ducts, which, as in the conifers, are in the bark and often the wood where they follow the rays and grain. Traumatic canals can be formed in the wood of some hardwoods, for example, sweet gum (*Liquidambar styraciflua*) and cherries.

Latex Latex (a milky mixture of such things as resins, oils, gums and proteins) is found in different plants from fungi to dandelions to trees. Many types of trees and shrubs have had their latex collected for making rubber (including the 'rubber tree', *Ficus elastica*, now grown commonly as a house plant). Around one third of rainforest trees have latex. The best commercial supply comes from *Hevea brasiliensis* in the spurge family (Euphorbiaceae) native to the Amazon and Orinoco river valleys of South America. The rich latex of this tree is about 33–75% water and 20–60% rubber. Latex is found in special ducts (lactifers) running through the bark in concentric circles. As with resin in conifers, the latex is under slight pressure, ensuring that any wound is sealed by coagulating latex (including those made deliberately to collect the latex, and which are treated with an anticoagulant to ensure a good collection).

Callus growth

The production of resins, gums and latex is often insufficient to seal over large wounds such as the breakage of large branches and the removal of areas of bark by, for example, squirrels. If the wound is kept artificially moist (by covering with plastic, lanolin or other non-toxic substance) so that the living cells of the rays do not dry out, a new bark will regenerate in the same season in many species. (Note, though, that in pruning off branches it is only the younger parenchyma cells of the sapwood around the edge of the stump that are capable of doing this, leaving a hole in the centre.) But if the newly exposed parenchyma cells are killed by toxic materials in paint or allowed to desiccate, then the wound can only be covered by the slow growth of callus tissue from the living cambium and rays around the edge. As the callus grows, it starts off uniformly around the wound but the sides tend to grow more rapidly (Figure 9.2) resulting in a circular wound ending up as a spindle-shaped scar (just why this is so is portrayed below under Wind). Once the sides of the callus meet, the cambium joins so new complete cylinders of wood are again laid down underneath the scarred bark.

Healing wounds

Unlike animals, trees cannot heal wounds; they can only cover them up. Once wood starts to rot it cannot be repaired. New wood can be grown over the top and look healthy but the rotting wood underneath is still there. Hence the wise saying quoted by William Pontey in *The Forest Pruner* (1810), 'An old oak is like a merchant, you never know his real worth till he be dead'.

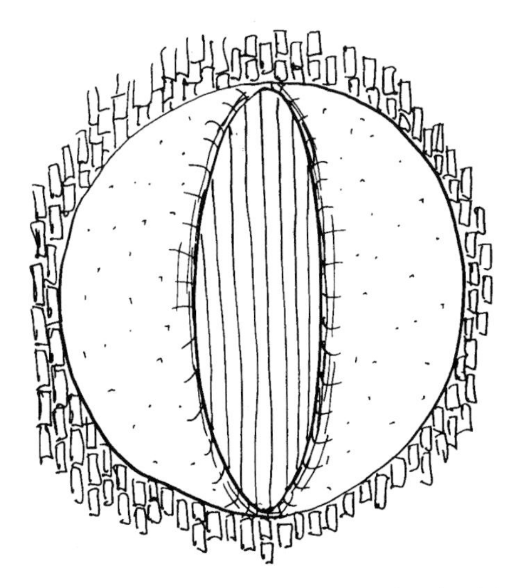

Figure 9.2. New callus growth over a wound is concentrated at the sides of the wound, producing an oval shape over the years, because this is where most stress is placed on the tree as it flexes in the wind. From: Mattheck, C. and Kubler, H. (1995). *Wood—The Internal Optimization of Trees.* Springer, Berlin, Figure 26a, Page 30.

Internal defences

Although resins, etc., and new wood help to seal off wounds there will always be some fungal spores and other damaging agents that slip by these first defences. So the tree must have other internal defences to deal with destructive agents once they get in. Before looking at those defences, it is worth looking at the biggest cause of the problem.

Fungal rot

If we are thinking of large structural decay, we are dealing primarily with fungal rots that are capable of decomposing the cell walls of wood for food. At an early stage of decay a number of fungi and bacteria will live on the nutritious cell contents but do little to structurally weaken wood. These are a nuisance to foresters since they cause staining of the wood, decreasing its value for timber and paper production. Structural decay comes from two groups of fungi. Brown rots attack the hemicellulose and cellulose of the wood leaving the lignin untouched as a brown mass usually cracked into cubes. White rots attack all components of the wood, reducing it to a light-coloured spongy mass.

Decay progresses most rapidly when the wood is moist, that is, above 20% moisture (furniture and timbers in a house are generally drier than this and hence don't rot; dry rot fungus gets around this by making its own water from sugars and by moving water considerable distances). Wood that is *too* wet is also safe from rot. Logs have traditionally been stored for long periods in ponds. And after the 1987 hurricane in southern England (when 15 million trees were brought down) some timber was stored in huge piles constantly wetted by sprinklers to be sold when prices had picked up. Wood has been stored in this way for up to 4 years with no appreciable deterioration in timber quality.

You may have noticed black lines running through a piece of well-rotted wood, especially in hardwoods, referred to by wood-turners as 'spalted wood'. These lines are made up of masses of gnarled dark-coloured fungal strands (hyphae) and are usually laid down by two different fungal species (or strains of one species) when they meet as a sort of garden fence between them (officially called a 'zone of antagonism'). Some fungi will, however, produce a black line around themselves even when no other fungus is present, perhaps marking where the fungus intermittently stops growing.

Defence of wood: sapwood and heartwood

As shown in Chapter 3, the wood inside a tree can be divided into a central core of heartwood surrounded by sapwood. Sapwood is responsible for conducting water and storing food in the living tissue of the rays. Heartwood is completely dead and filled with chemicals designed to repel all boarders. Over time the innermost sapwood is gradually converted into heartwood. The creation of heartwood is a deliberate process and is not, as is sometimes said, just old living tissue gradually fading away. It has also been argued that heartwood is a very useful dumping ground for waste products that would otherwise be hard to get rid of. Against this must be set the knowledge that the compounds incorporated into heartwood are specifically and expensively produced. The energy cost per unit mass of these compounds is twice that of wood.

One of the first processes towards the formation of heartwood is to block the water-carrying tubes, although this is primarily in response to air in the system and may happen some time before heartwood is formed. In conifers the pits close (see Chapter 3) when air enters the tubes, and resins may add further blockage. In hardwoods, where living cells in sapwood are more numerous, vessels can be plugged by the cells either exuding gums or producing balloon-like outgrowths (called tyloses) into the vessel (Figure 9.3): the original air-bags! Trees with small vessels tend to have gum plugs whereas those with larger vessels have tyloses (although some trees have both). This plugging of cells starts in the sapwood as the tube becomes air-filled: again, see Chapter 3 for further explanation.

Blockage of the tubes by closing of pits and adding gums and tyloses is normally complete by the time the heartwood is formed, but not all trees do block up the tubes: red oak—*Quercus rubra*—has very few tyloses and it is possible to blow through a piece of heartwood (which also explains why it is useless for making barrels!). These blockages will inevitably slow the movement of rot through

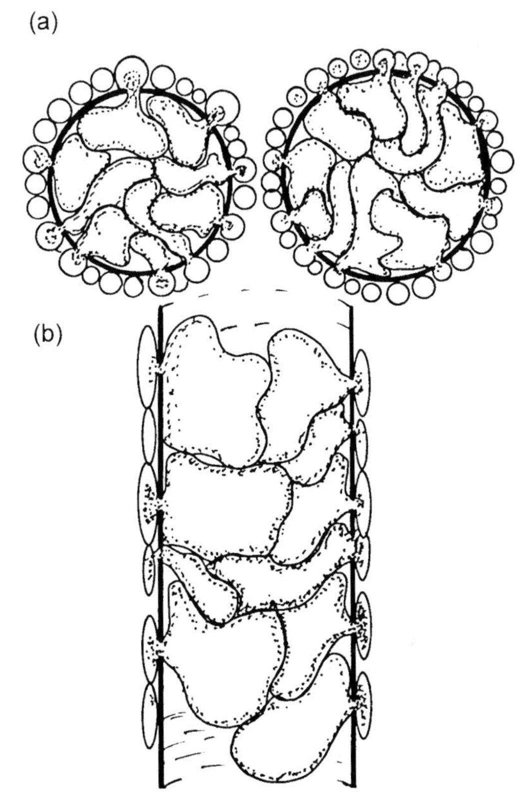

Figure 9.3. The balloon-like outgrowths (tyloses) found in some hardwoods, which burst into the water-conducting tubes of the wood from surrounding living cells to form a plug: (a) looking from above and (b) looking from the side. From: Shigo, A.L. (1991). *Modern Arboriculture.* Shigo and Trees, Durham, NH.

the wood since it must digest its way through rather than growing up unhindered tubes.

As sapwood is turned into heartwood, the living cells on the edge of the heartwood die and any food they are storing is either used in making the heartwood or moved elsewhere. Cell death is accompanied by the formation of a variety of compounds (typically lignins, polyphenols, gums and resins) known collectively as 'extractables'. These include some compounds that we find pleasurable but which are toxic to wood-deteriorating organisms. Think of camphor-wood (*Cinnamomum camphora*), used to make chests that repel insects, sandalwood (*Santalum album*), used for centuries as incense in the Orient, and the distinctive oils in cedar wood (*Cedrus* spp.). Others are less inviting. Volatile oils give fresh elm a distinctive and unpleasant aroma; my colleagues once complained that something had died in the service ducts when I was storing a fresh elm slice in my office!

The extractives are added to both the walls and the bore of the former water-conducting tubes and account for the greater density and durability of heartwood. Woods with few extractives (e.g. alder, ash, beech, lime, willow) may rot in less than 5 years when in contact with the ground, whereas extractive-rich woods (e.g. oak, sweet chestnut, western red cedar and yew) are usually durable for more than 25 years. Having said this, elm is usually considered non-durable and indeed makes short-lived fence posts, but keep it wet and it lasts for hundreds of years. The Romans knew this and used hollow elm logs as water pipes (apparently it was a long 'boring' job drilling these out, hence our use of this term). Elm water pipes unearthed in 1930 in London were still sound after more then 300 years underground.

Compartmentalisation of decay

Decay-resistant heartwood goes a long way towards keeping a tree full of wood but wood is not just a lump of hard-to-decay material; protection is much more subtle. Alex Shigo, an American forester who has spent decades looking at the insides of trees, has championed the idea that wood is divided into compartments. He saw that rot does not always spread through the whole tree even when it has had sufficient time. There appear to be three 'walls' preventing the spread of disease (Figure 9.4a). Wall 1 resists vertical decay and not surprisingly, given the structure of wood, is the weakest, allowing fairly rapid spread of rot up and down the tree; tyloses and gums presumably play a part in this 'wall' since wood without them tends to be least durable. Wall 2, corresponding to the ring boundaries, resists inward decay, and Wall 3 (the rays) radial spread. These chemical walls are laid down as the wood is formed, in a similar manner to the way a regular grid of anti-tank road blocks might be laid down in a military retreat.

After a tree is wounded by, for example, being hit by a lawn-mower, these three walls will act to slow down the spread of rot that enters. But that is not all. In the next growing season after the damage, the cambium lays down a fourth wall reinforced by resins and other defensive compounds—called the barrier zone—between the wood present at the time of wounding and the wood grown afterwards (Figure 9.4b). The barrier zone may stretch right around the tree or be restricted to a small arc around the circumference, particularly at the wound site, like a sticky plaster on a child's knee. This reaction after the damage has occurred provides the strongest wall yet to isolate the rot from the new wood. Strong though it is, the wall is not infallible and rot can eventually break through. If this happens, the tree responds by laying down another barrier zone. The cambium may not be touched by the breakaway rot but it receives the message (via the living tissue in the rays?). The spread of rot can be sufficiently fast that a barrier zone does not have time to form before the trunk is killed at that point.

The value of this barrier is that a tree can be suffering from rot of the heart-

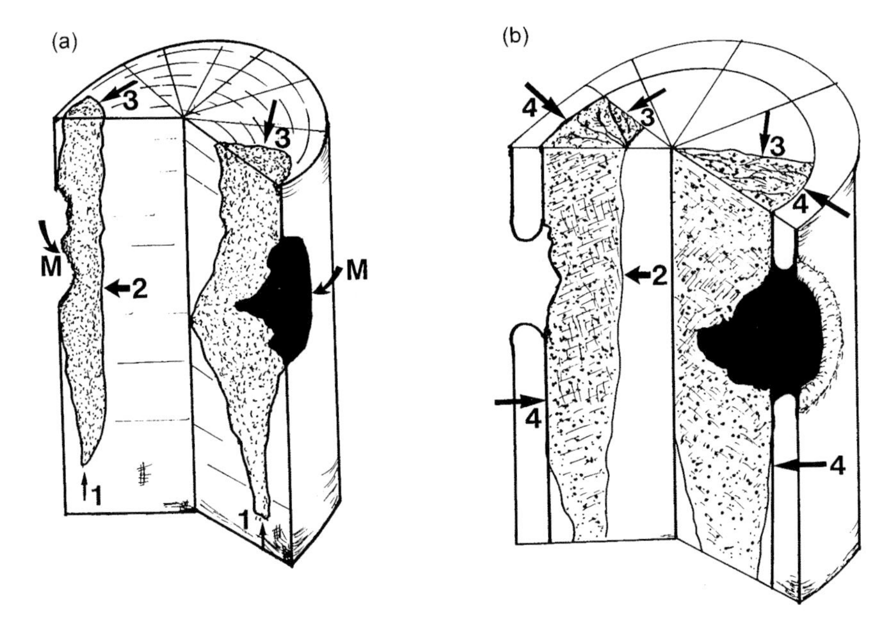

Figure 9.4. Compartmentalization of decay in wood. (a) When a tree is wounded (marked M) three walls (1–3) prevent the spread of rot vertically, towards the centre and radially. These correspond to the tube cross-walls, the growth rings and rays, respectively. (b) New growth after the wounding is isolated from the damaged centre by the formation of a strong fourth wall—the 'barrier zone'—(marked 4). As the rot spreads past walls 1–3 to fill the centre of the tree, it does not readily penetrate through to the new wood. Thus a tree with rot in the centre is not necessarily immediately doomed. From: Shigo, A.L. (1991). *Modern Arboriculture.* Shigo and Trees, Durham, NH.

wood and become hollow without being compromised and imminently doomed; the tree is compartmentalised to protect the outer living skin of the tree. Of course, it may be so rotten that it physically snaps: this will be looked at later in the chapter.

Damage from the environment

Harsh conditions and pollution

The environment can throw up any number of problems—extreme temperatures, drought, flooding, lightning (see Chapter 3), soil impoverishment, wind, fire—and the list goes on. We are also merrily adding to these problems by simple things such as deep ploughing (which damages roots), extensive use of herbicides and large-scale pollution; this list could also go on.

Pollution is a perennial concern. Fortunately, a number of trees are relatively tolerant. London plane (*Platanus* × *hispanica*) is particularly tolerant of grimy city atmospheres with a noted ability to root in compacted and covered soil. A number of trees are relatively resistant to sulphur dioxide including Lawson cypress (*Chamaecyparis lawsoniana*), junipers (*Juniperus* spp.), Corsican pine (*Pinus nigra*), western red cedar (*Thuja plicata*), ginkgo (*Ginkgo biloba*), beech, oaks, hornbeam (*Carpinus betulus*) and planes (*Platanus* spp.). Nevertheless, there is plenty of evidence that pollutants do affect trees.

Acid rain kills roots and can adversely affect mycorrhizas through soil acidification, and it can strip large amounts of nutrients from the leaves as it falls through the canopy.
Excess nitrogen from burning fossil fuels and too many cattle (which release huge quantities of ammonia) has led to soil acidification and potential forest decline in the Netherlands.
Too much ozone disrupts photosynthesis and, with other pollutants, can sterilize pollen and so reduce seed production. Too little overall in the upper atmosphere will expose trees to unhealthy quantities of ultraviolet light.
Heavy metal contamination can be directly lethal and indirectly so through changing soil acidity.
Particles of soot on leaves can also be harmful by shading. A 2500 year old fig tree in north India under which Buddha attained enlightenment is under threat from smoke from pilgrims' candles: soot is blackening the leaves and preventing photosynthesis.

In urban areas you can add a whole gamut of extra stresses.

De-icing salt on roads.
Tarmac over roots.
Weed and grass competition.
Untold quantities of dog urine (which provides excess potassium) and faeces.
Soil compaction (Chapter 2).
Vandalism.

Is it any wonder that urban trees have a reputation for being short-lived? This is especially true of isolated trees, such as those in hedgerows, which are probably more prone to damage because they filter out more pollution than woodland trees, which protect each other. In New York, the average lifespan of trees is 7–40 years.

Cold

We saw in Chapter 3 that cold weather can cause frost cracks in trunks. Cold can also cause the frost heaving of young trees and death of small twigs. The latter is especially a problem in springs with late frosts, which catch the buds as they lose their cold-hardiness (walnuts, ashes, sweet chestnut, oaks and beech are vulnerable; birches, hazel, hornbeam, limes, elms and many poplars are more tolerant). Evergreens are not immune. The foliage of the 'Elegans' variety of Japanese red cedar (*Cryptomeria japonica* 'Elegans') and young western red cedars (*Thuja plicata*), among others, turns a most delicious bronze over winter. More severe is the 'red belt' damage that occurs in conifers when the air is warm while the ground is still frozen: the leaves lose too much water and die.

Despite such damage, many trees can survive low temperatures given the right 'hardening off'. A variety of mechanisms give different degrees of protection. Plants in fairly mild climates can survive purely by the insulation given by bark (like lagging on pipes). If living cells approach freezing, the first step is to prevent ice forming in cells because the crystals would puncture the cells and kill them. The first line of defence is depression of the freezing point by accumulating sugars, organic acids or amino acids inside the cell, usually in response to short day length (e.g. white spruce, *Picea glauca*) or lowered temperatures (e.g. Nootka cypress, *Chamaecyparis nootkatensis*, and western red cedar, *Thuja plicata*). This works in the same way that adding salt to roads prevents them freezing. Using this mechanism plants tolerate mild frosts down to –1 or –2 °C. The second line of defence common in woody plants is provided by the cell contents supercooling without freezing. This gives protection down to as low as –40 °C but needs to be preceded by several days below 5 °C (which explains why a cold spell in early winter can kill when the same temperature later does not, and why rapid fluctuations between freezing and thawing, as in alpine areas, can damage even the most hardy of trees). The third line of defence, needed in only the most severe climates by trees such as some birches, poplars and willows, is to take all the freezable water from the cells and allow it to freeze *between* the cells while tolerating the extreme desiccation caused inside the cells. This appears to be conditional on slow continual cooling but if these conditions are met it allows dormant twigs of willow and many other northern woody plants to survive down to at least the temperature of liquid nitrogen (–196 °C) without harm!

Heat and fire

In temperate regions, temperatures much above 17–20 °C slow down photosynthesis and once they reach 50 °C most living tissue is well on the way to dying. Under normal circumstances, a plant has a number of ways of staying cool with evaporation of water being foremost. Plants can cope with most environments, even deserts where water is in short supply. But how can plants cope with temperatures in a fire that are in the order of 800–1000 °C whether we are talking about a candle flame or a raging forest fire?

The first point to appreciate is that fire is a naturally occurring event in many types of vegetation, including forests, around the world. A patch of prairie grassland may burn every 1–10 years and dry deciduous and conifer forests around the globe would naturally be burnt every 50–150 years on average. Moist broadleaved forests tend to be less flammable but even seasonal tropical rainforest (where dry periods occur regularly) will burn every few centuries. Because fires have been burning around the world for perhaps 10 million years (caused mainly by lightning), we should expect that plants will not only be able to cope with high temperatures but even be able to take advantage of the situation. And that's exactly what you find.

Many trees, like aspen, have thin bark and are easily killed by even gentle fires. But the soil is a good insulator and temperatures above 100 °C are rarely found below 2 cm. The roots of aspen and many other woodland shrubs and herbs thus survive to resprout. Other trees have different strategies. Douglas fir (*Pseudotsuga menziesii*) and the giant sequoia (*Sequoiadendron giganteum*) of western N America, among others, have developed thick non-flammable bark to insulate the living tissue from the heat of the flames. In the Sierra Nevada mountains of California, fires would naturally burn through the giant sequoia forest at something of the order of every 10 years. Since the giant sequoias live for around 2000 years they will meet maybe 200 fires in their life. Because the fires are so frequent, comparatively little burnable material (dead needles, twigs, etc.) builds up, so the fires are comparatively gentle. Add to this a bark that is thick (often 30 cm thick and up to 80 cm), full of insulating air, and virtually non-flammable, and you can see why the trees can survive so many fires. By contrast its thin-skinned competitors, such as white fir (*Abies concolor*), are more readily killed. Fire thus allows the giant sequoias to rule the forest. Heat-proof bark is also seen to a lesser extent in temperate deciduous forests, where thin-barked American beech (*Fagus grandifolia*) and American plane (*Platanus occidentalis*) are more readily killed than the thicker-barked oaks and American chestnut (*Castanea dentata*).

The giant sequoia has another trick up its metaphorical sleeves. Along with half a dozen pines and many other trees around the world (see page 216), the giant sequoia keeps its cones tightly shut for decades, a condition known as serotiny. In this way thousands of seeds are stored in the canopy for decades (pines can hold cones with viable seed for 20–50 years or more, *Banksia* species of Australia for 10–20 years, *Protea* species of S Africa for just a few years). The seeds are most commonly released by fire.

In serotinous conifers the cone scales are sealed shut with resin and few if any seeds are released until the cones are heated to 50–60 °C. As nature would have it, this sort of temperature in the canopy is likely to be reached only in those fires hot enough to kill the parents. Within hours or a few days of the fire, the scales open in the daytime heat and a flurry of seed drops to the ground, which by then is perfectly cool. In the southern hemisphere the seeds are equally well protected by woody fruits: the banksias have woody fruits (follicles) moulded together by woody bracts into the characteristic 'bottle', and the she-oak has a woody cone-like fruit. Characteristically in banksias the old flowers do not fall off and the heat produced by these catching fire (100–300 °C for a few minutes) is enough to induce opening. None of these serotinous fruits is remarkably heat-resistant but they are well capable of protecting the seeds from the brief high temperatures encountered.

It just so happens that the seeds of serotinous species do best on a burnt soil (lack of competition, plenty of light, warmth and abundant nutrients in the ash). Thus over the next few years the forest becomes dominated by the next generation of the serotinous species. Like the phoenix, they emerge from the ashes of their own species. So good is the mechanism that many serotinous species have evolved to be highly flammable with peeling bark (as in eucalypts) or dead branches (as in pines) to carry the fire into the canopy, aiding complete immolation of self and neighbours. The other side of the coin can be currently seen in Canada's forests where fire-prone areas are naturally dominated by fire-adapted jack pine (*Pinus banksiana*) and lodgepole pine (*P. contorta*). Where fires have been suppressed, the fire-created stands are old and frail and a new vigorous layer of fire-intolerant firs and hemlocks grow up underneath to take the pines' place.

A number of trees regenerate after fire by storing their seed in the soil (Chapter 8). The trick is to detect when the fire has burnt past. In some cases the fire itself is the trigger. Gorse seeds (*Ulex* spp.) in European heathlands have a hard seed coat that is split by the high fire temperatures so allowing in water and thus germination. Others (e.g. heathers) detect not the fire but the fluctuating temperatures that result from the removal of the vegetation. A number of South African and Australian species are stimulated to germinate by chemicals in the smoke, and shrubs in the Californian chaparral are stimulated by the *removal* of inhibiting chemicals (put out by nasty neighbours) by the fire.

The clear advantage of seed storage in the canopy is the mass release of seed after a fire, which takes advantage of a good seedbed cleared of debris and competition, and satiates predators in the same way as masting (Chapter 5), producing a dense growth of the next generation. But why store seed in the canopy rather than in the soil? Serotiny appears to put all seeds into one post-fire basket, and if the growing season following a fire is disastrous then everything is lost. The first part of the answer is that the seed supply is not as one-off as it sounds. Most serotinous species leak out a few seed over the years in between fires, and once seeds are released by fire a small proportion usually do not germinate till the following year or so, giving a small insurance policy against disastrous years. The second part of the answer is that fires can burn off the top few centimetres of litter and organic debris which contain the bulk of the soil seed bank. So it is safer to store seeds in the canopy, providing a fire is likely to come along before the tree dies and the stored seed is lost.

Nevertheless, not all species in fire-prone forests are serotinous, and their abundance has a lot to do with how often a fire is expected in relation to the reproductive lifespan of a tree. If fires tend to be very frequent (every decade or less) or unpredictable, then trees that are tolerant of burning (by having thick bark) or are good at resprouting from the burnt stump are likely to dominate. When the normal interval between fires exceeds the lifespan of a tree, serotiny (canopy storage of seeds) will be of little use since once the tree dies the stored seed will not survive for long. In this case the next generation must come from seed stored in the soil or seed coming in from outside the area. In between these extremes lies the realm of serotinous species.

There are many other adaptations to fire which lack of room does not permit to mention. But I cannot pass without mentioning the longleaf pine (*Pinus palustris*) of the southeast USA. It has a problem in that the young seedlings are readily killed by the frequent fires that sweep through its open grassy habitat. If they can survive to reach 2 m in height, the bark is thick enough to resist fire and they are comparatively safe. To solve the problem, the newly germinated seedling goes through a 'grass-stage' where the shoot grows to several centimetres in a few weeks and then stops growing above ground for 3–7 years. During this time the bud is protected from fire by a dense tuft of needles, and all energy goes into the roots. When sufficient energy has been stored the shoot makes a mad dash upwards, growing rapidly to reduce the time spent as a vulnerable juvenile.

A final note on firewood. Ash (*Fraxinus* spp.) and holly (*Ilex aquifolium*) are renowned for making good firewood, which will burn even when green. Is this related to their fire ecology? On the whole, no: these species rarely meet fire naturally. Rather it is the nature of their storing food as oils (Chapter 2). In ash it is the large amounts of oleic acid (a fatty acid constituent of olive oil) that makes the wood so flammable.

Wind

Whole woodlands of trees broken off or uprooted gives a lasting impression of wind as a devastating agent that the tree can do little to defend itself against. Despite appearances, trees have a variety of mechanisms for coping with wind. Wind blowing against a tree acts like a hand pushing on the end of a lever acting to bend the tree (in the same way that a heavy branch or leaning tree bends under the pull of gravity, as discussed in Chapter 7). The taller the tree, the stronger the lever pushing against the tree (in the same way that a longer spanner more easily shifts a tight bolt). Trees overcome this to a certain extent by being flexible. When the wind blows, the branches bend and, in effect, reduce the height of the tree by 'putting its ears back'. As wind speed increases, smaller branches are snapped as a sacrifice to save the trunk by further reducing the sail area and height of the tree. The shorter the apparent height, the less force exerted by the wind at the base.

The shape of the tree is also carefully adjusted to reduce weak areas where the tree might snap. As a tree is bent by the wind it will snap if one area has more

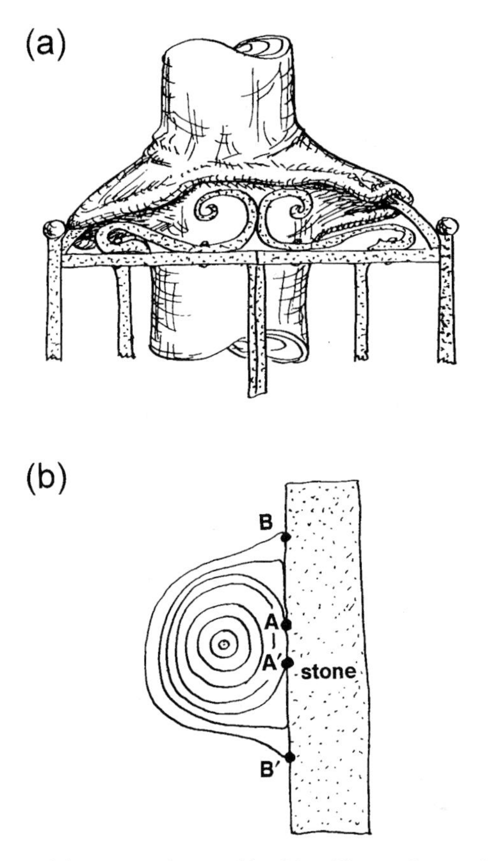

Figure 9.5. When a tree (a) meets an immovable object like a railing or (b) has something like a stone leant against it, the tree will detect the stress at that point and grow extra wood to spread the load over as large an area as possible to reduce the risk of snapping at that point. In (b) the original contact zone A–A is made longer (B–B). From: Mattheck, C. and Kubler, H. (1995). *Wood—The Internal Optimization of Trees.* Springer, Berlin, Figures 29b & 30, Pages 23 & 24.

stress on it than others. But the cambium can detect stressed areas and lays down more wood to strengthen them. Conversely, areas experiencing less stress will have less wood expended on them. This is Mattheck's axiom of uniform stress: an optimal structure has a uniform stress over the whole of its surface. As the tree flexes in the wind, most stress is felt in the outer skin of the tree. In the same way that a rock in a river forces the water aside and increases the water speed at its sides, so a defect in a tree's surface will divert the stress lines around itself such that the stress is greatest at the sides of the defect. Any cracking of the trunk will start at the sides of a hole, and this is where 'repair' of damage is concentrated. The tree will detect the stress and concentrate new callus growth at the sides of the hole (see Figure 9.2). In the same way, when a heavy object is leant against a tree, or a tree getting fatter meets an immovable object like a railing, the tree detects the stress and adds more wood at this point to spread the load over as large an area as possible to reduce the stress on any one point (Figure 9.5). This is analogous to leaning against the corner of a table and putting a cushion under your bottom to spread the load more evenly.

Trees still fail: why?

Wood is expensive stuff to produce. A tree that puts wood where it is not needed will be wasting resources and will likely suffer in the competitive struggle for life. So the design of trees, like most things, is a compromise between risk and safety. Trees are not so overdesigned as to protect against any eventuality. A natural consequence of this is that trees will sometimes break when under extreme stress such as hurricanes. So, if a tree falls over in high winds and flattens your car you can't necessarily blame or sue someone, you have to accept that this is how trees are designed. And is this so unacceptable? High winds blow down chimneys and cause roofs to shed tiles, yet we don't banish people's roofs and chimneys in the same way we do trees!

Having said this, however, some trees are safer than others because they evolve different levels of compromise between risk and safety. A tree like an oak has a high investment in abundant, strong and well-defended wood and so for its height is more likely to withstand extreme stresses and live to a ripe old age. The negative side is that it will be less competitive and is likely to suffer more extreme competition, especially when young. Birch, as a pioneer invading open areas, works with a lower safety margin, putting more emphasis on quick height growth with a tall thin trunk of less durable wood, and rapid reproduction. With that goes the greater risk of catastrophe and early death: it's a boom and bust strategy. Moreover, within a species, individuals will 'decide' on their safety margin according to need. Trees in shade are particularly vulnerable to damage because of their attempt to reach light with a consequent neglect of mechanical strength.

Hollow trees

Whatever the strategy of safety used by a tree, it is intuitively right that rot eating away the centre of a tree will make it more prone to being snapped in the wind. Or is it? Claus Mattheck looked at 1200 trees of a variety of species, some that had snapped, others that hadn't, and measured the amount of sound wood left around the edge in proportion to the width of the tree. He found that regardless of species or size, trees rarely snapped until more than 70% of the diameter was rotted away (i.e. the wall thickness was less than 0.3 of the radius). Thus a tree 50 cm in diameter (i.e. 25 cm radius), should have walls at least 7.5 cm thick (0.3×25 cm) to ensure safety, and a tree 100 cm in diameter should have walls 15 cm thick. Trees with thinner walls than the above can be found standing;

Figure 9.6. A felled Huntingdon elm (*Ulmus* × *hollandica* 'Vegeta') showing the pencil-like roots growing down the hollow centre. The fine feeding roots would have been absorbing nutrients from the decomposing wood. An example of self-recycling!

these generally have a small canopy and so are less affected by wind. This may seem like a remarkably small amount of sound wood to hold up a large tree, which underlines how well designed wood and trees are.

These results clearly show that much of the wood in a tree trunk is redundant when it comes to holding the tree up. Indeed, hollownesss can be an advantage. After severe gales (such as those in southern England in 1987) many hollow trees were left standing while their solid neighbours fell. This may be partly explained by old hollow trees having a smaller canopy—a smaller sail area—but it seems that hollow trees are more flexible and better able to withstand the buffeting and swaying.

Hollowness is beneficial in another way. It is well known among tropical trees, but also temperate trees such as yews and elms, that roots can be produced from the sound wood (adventitious roots) into the rotting mass in the centre. This allows the tree to suck up the nutrients being released by the rotting and, in effect, recycle itself. What's more, any droppings or dead remains left by animals living in the hollow add to the soup, all of which gives the tree extra supplies in its battle for dominance. An elm I saw felled not so long ago had several dozen pencil-thickness roots growing down the centre with fine feeding roots firmly embedded in the brown rot around the inside of the hollow (Figure 9.6).

So when you see a hollow tree don't feel sorry; hollowness appears to be a deliberate part of its survival strategy, making it more flexible and better able to resist high winds, and allowing it to recycle itself!

Windthrow: uprooting trees

As explained in Chapter 3, it is the big framework roots (which spread roughly the width of the canopy) that hold the tree up. These roots create a 'root plate' of soil welded together by roots although in most trees (which have a tap root and sinkers) this may be more aptly described as a 'root ball'.

How does a tree fall over? As a tree sways in the wind the soil begins to crack and horizontal roots are pulled out or snapped on the windward side (Figure 9.7a,b) . As these 'guy roots' are removed, the root ball begins to rotate in its socket of soil and the tree starts to lean, tearing out more roots as it goes (Figure 9.7c). The tree forms less of a target for the wind but its weight, now being way off-centre, helps to bring the tree crashing down (Figure 9.7d). It is the friction between the root ball and the surrounding soil that really holds the tree up. Thus wet soils (where the water reduces friction) are likely to result in more windthrow. Another consequence of this friction is that as a tree gets bigger and heavier (pushing more into the soil and increasing the friction) the root plate is proportionately smaller: small trees may have a root plate more than 15 times the stem radius whereas in larger trees it need to be no more than three times the stem radius.

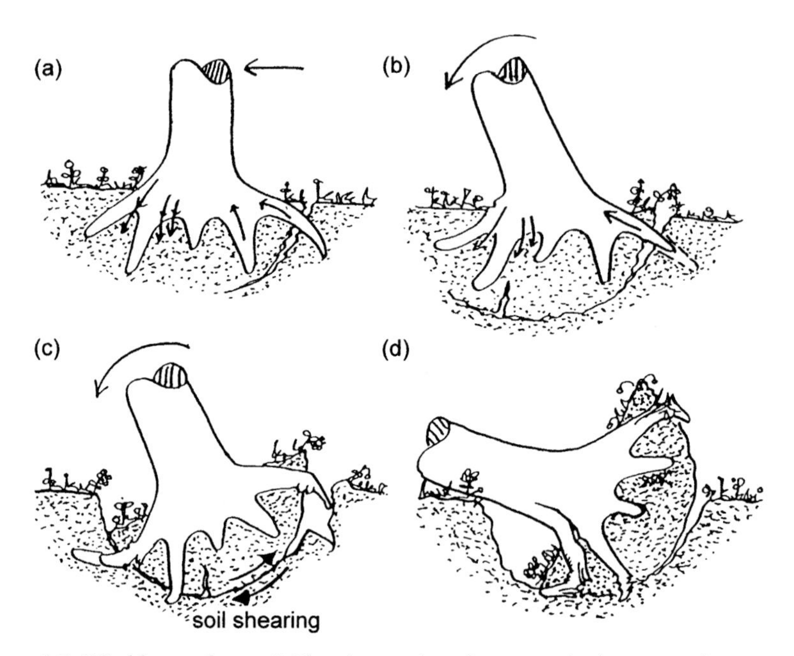

Figure 9.7. Windthrow of a tree I. The pictures show four stages in the process for a tree with comparatively deep roots where the tree falls by the root ball rotating in an earthen socket. After Coutts from Mattheck, C. and Breloer, H. (1994). *The Body Language of Trees: a Handbook for Failure Analysis.* Research for Amenity Trees, No. 4, HMSO, London.

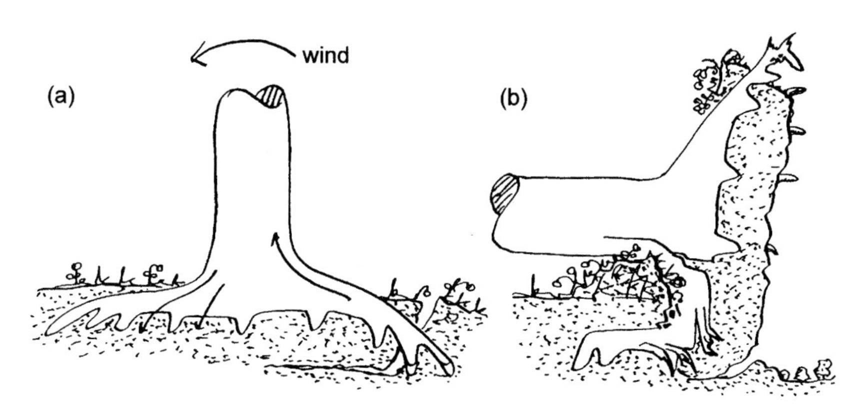

Figure 9.8. Windthrow of a tree II. A tree with shallow roots falls in the same way that a coat stand topples over. The weight of the tree and root plate must be lifted as it pivots over the edge of the root plate. From: Mattheck, C. and Breloer, H. (1994). *The Body Language of Trees: a Handbook for Failure Analysis.* Research for Amenity Trees, No. 4, HMSO, London.

Trees with very shallow roots—such as spruce and beech—fall in a slightly different way. The initial stages are the same as above with the soil cracking and windward roots being pulled out or broken (Figure 9.8a). But then the whole root plate hinges along the leeward side and falls, just like a hatstand toppling over (Figure 9.8b). What holds these trees up is firstly the weight of the tree and the soil carried on the root plate, which must be lifted, and secondly the length of the roots (longer roots make the lever from the hinge to the centre of gravity longer and so harder work). Friction has little to do with the process. Thus the shallow roots that are inevitable on wet soils are the best type for holding the trees up! Nevertheless, the environmental dice are loaded against the tree, and shallow-rooted trees are still less stable than deeper-rooted neighbours. Windthrow is common among conifer plantations in upland Britain for this reason.

Trees are not passive in the face of wind. They usually grow more roots on the side towards prevailing winds and to a lesser extent on the lee side, with fewest at right angles to the wind. Winds are usually highest in winter when the trees are mostly dormant but there is evidence from Scots pine (*Pinus sylvestris*) that trees can retain information about mechanical forces acting on their stems during the winter and respond the following growing season (see Lundqvist and Valinger 1996). As discussed in the sections above, trees do not squander limited resources; they do not grow roots that are not needed. Thus, trees in the middle of groups often have less well developed roots compared with the individuals around the edges. This represents good economy until part of the stand is felled or blown over and the remainder is exposed to high winds before extra roots can be grown.

Does wind kill trees?

It is now (finally) recognised that a large proportion of the 15 million trees that snapped and blew over in the 1987 gales in southeast England were not as dead as they first looked. A tree left leaning after high winds is by no means finished. The roots regrow and the tree straightens itself by growing 'reaction wood' (Chapter 3): too many leaning trees are given up as lost and cleared away by tidy-minded people. Even a completely prostrate tree with some roots intact (more likely in a shallow-rooted tree; Figure 9.8) can produce suckers from the surviving roots or create new trunks out of branches now pointing skywards (and so creating strange lines of trees to fox the unwary naturalist once the original stem is buried in leaf litter!). Similarly, the stump of a snapped tree can usually grow new shoots from stored (epicormic) buds. The exception, where winds *can* kill trees, is found in old weak trees, as described at the end of this chapter.

The age of trees

Despite all the factors working against trees surviving, they are long-lived by the standards of most plants and animals (hence the Bible passage in Isaiah 65:22 'As the days of a tree are the days of my people'), and, of course, the oldest known living things in the world are trees! The life span of most trees is measured in centuries, most being comfortably seen out in less than 500 years (see Box 9.1). Tropical trees are probably no exception (but given the normal lack of annual rings it is hard to tell). Richards, in his acclaimed book *The Tropical Rainforest*, says that the average maximum lifespan of only two species is known: *Shorea leprosula* at 250 years and *Parashorea malaanonan* at 200 years with a maximum age of perhaps 300–350 years. Speculation that some species may be capable of living to almost 1500 years is just that. Work on the revegetation of Krakatoa suggests the average lifespan of tropical trees there to be 80–120 years.

Around the world there are a number of trees that live significantly longer. Many of the redwoods are long-lived, with the coastal redwood (*Sequoia sempervirens*) and giant sequoia (*Sequoiadendron giganteum*) living over two and three thousand years, respectively. The oldest living trunks belong to the bristlecone pines (*Pinus aristata*[1]) growing at more than 3000 m (10 000 ft) in California and Nevada. The oldest known living specimen has been aged by counting the rings to be more than 4600 years old. An older but scrawnier specimen, 4900 years old, was inadvertently cut down to count the rings, and a life span of over 5000 years seems possible. When the Egyptians were building the pyramids these trees were already well established.

[1] There is some difference of opinion over the correct Latin name. Americans like to call these old bristlecone pines *Pinus longaeva*. The *World Checklist of Conifers* (1993), however, calls them *Pinus aristata* var. *longaeva*.

Box 9.1. The expected life span of a range of trees in years

Birches	*Betula* spp.	80–200
Red maple	*Acer rubrum*	110
Ash	*Fraxinus excelsior*	200–300
Beech	*Fagus sylvatica*	200–400
Scots pine	*Pinus sylvestris*	500
Oak	*Quercus* spp.	700–1000
Douglas fir	*Pseudotsuga menziesii*	750
Ponderosa pine	*Pinus ponderosa*	1000+
Limber pine	*Pinus flexilis*	2000
Coastal redwood	*Sequoia sempervirens*	2000+
Giant sequoia	*Sequoiadendron giganteum*	3000+
Bristlecone pine	*Pinus aristata*	4900+
Yew	*Taxus baccata*	possibly 5000+

Other types of tree *may* be even older. The dragon trees (*Dracaena draco*) on Sumatra are thought by some to be up to 10 000 years old, and some cycads are speculated to be 14 000 years or older. Since neither of these produce annual rings or have a solid centre that can be carbon dated, I suspect that these estimates are tales that have grown in fond telling.

Stories occasionally reach the newspapers of incredibly old trees being discovered. For example, a creosote bush (*Larrea tridentata*) in California has been dated as 11 700 years old, a Huon pine (*Lagarostrobus franklinii*) in Tasmania is possibly 10 000 years old, and a shrub called King's holly (*Lomatia tasmania*) growing in southwest Tasmania may be up to 40 000 years old. Are these for real? Well, yes and no. In these cases it is the plant clump that is so ancient. The creosote bush was dated by looking at the size of the fairy-ring-like circles, which grow outwards at a known rate. No one stem is anywhere near the age of the clump. Similarly, the Huon pine and King's holly are clones, which have probably been resident on the site for thousands of years but the current stems are all fairly young. In the Huon pine there is no trunk older than around 2000 years. It is rather like having an old broom that has done fifty years' service with only five new heads and two new handles. Whether you wish to count the bristlecone pines or King's holly as the oldest plant is a debatable point: both are equally impressive. Personally, to touch a bristlecone pine trunk that has been growing for over four thousand years takes some beating.

Yew trees in Britain have always been thought of as well capable of living for a number of centuries but it is now appearing that they can live for thousands of years. Certainly the largest are impressive. The crumbling shell of the Fortingall Yew on Tayside, Scotland, has a staggering girth (circumference) of around 17 m and is suggested to be up to 5000 years old. A stumbling block in deciding precisely *how* old yews are has been the mechanics of ageing them. The oldest specimens are hollow, and even if they were not, it would be difficult to get access to the rings. Felling a venerable tree is a little drastic just to see how old it is. A special corer can be used to extract a thin straw of wood from the trunk but is no easy matter in a tree several metres in diameter! Mitchell's rule (see Chapter 6) of 1/2 to 1 inch growth in girth per year can be a useful way of estimating age in a non-destructive way. This can be made more accurate by using girth data from trees where the rings *have* been counted or the planting date is known to construct a graph of girth against age (see, for example, Stephenson and Demetry 1995 but also White 1998). Yews, however, do not seem to necessarily put on annual rings (see Chapter 3) and trees are known that show no growth in girth for over 300 years (though loss of part of the trunk or different accuracies/heights of measuring need to be considered). Repeated measurements have shown that the average growth in girth of yews may be just 5 mm per year, reducing to 0.3 mm per year in old trees (corresponding to ring-widths of 0.8 mm and 0.05 mm, respectively). This is much slower than Mitchell's 1/2 to 1 inch, so the Fortingall Yew *could* be 5,000 years old. Perhaps the most convincing evidence for the antiquity of yews is from a tree at Tandridge in Surrey, which is next to a church whose Saxon foundations were built around its roots. Roots grow fatter very slowly so even a thousand years ago the tree must have been quite sizable to have roots big enough to be bridged. And yet at 11 m in girth it is by no means the largest tree in the country, and is estimated to be a mere 2500 years old. The unexpected age of yews is perhaps the answer to the age-old question of why yews are traditionally planted in churchyards: the churches were built around the yews revered by pagans.

What kills a tree?

The tree's environment

A tree may readily succumb to some deficiency (such as drought) or excess (e.g. frost, salt or pollution) in its environment; see Box 9.2. Although we know at least partly the effects of individual pollutants and other stresses on the physiology and performance of trees, it is often difficult to pinpoint the precise cause behind a sick tree or forest. This is sometimes because problems share the same symptoms. For example, natural gas leaking from underground pipelines is not directly poisonous but displaces oxygen and produces similar symptoms to waterlogging or soil compaction. Moreover, oxidation of natural gas (methane) by bacteria produces water, so gas-leak sites are often wetter! To make life more complicated the lethal fungal disease phytophthora is closely associated with wet soil conditions. So what *appears* to be killing the tree may be masking a much more deadly cause. Trees weakened by one thing, such as disease or pollution, are more susceptible to another, such as drought, windthrow or another disease (see page 165). The obvious cause of death then gets the blame, letting the initial cause of decline off the hook. In a similar way, small stresses can add up to one serious problem. 'Forest decline' which 'has become an ecological crisis throughout the developed world' (Klein and Perkins 1988) appears to have no single cause but to be an amalgam of a number of stresses.

Virulent diseases

Sudden death can come with diseases that are able to rapidly and catastrophically overcome the defences of the infected tree. Honey fungus is probably the most notorious (Box 9.2). A number of other diseases are specific to certain types of tree and have caused consternation around the world. Chestnut blight in N America and Dutch elm disease come readily to mind although a number of others could be listed, and a number are undoubtedly waiting in the wings.

Dutch elm disease Dutch elm disease is worth looking at as an exemplary virulent disease. The fungus behind the disease, first investigated by the Dutch,

Box 9.2. Susceptibility of trees to different problems

These are not complete lists; refer to Philips and Burdekin (1982) or a good gardening manual for further information.

Frost

Susceptible		Tolerant	
Walnuts and other *Juglans* species		Birches	*Betula* spp.
Ash	*Fraxinus excelsior*	Hazel	*Corylus avellana*
Sweet chestnut	*Castanea sativa*	Hornbeam	*Carpinus betulus*
Oaks	*Quercus* spp.	Limes	*Tilia* spp.
Beech	*Fagus sylvatica*	Elms	*Ulmus* spp.
Grand fir	*Abies grandis*	Many Poplars	*Populus* spp.
Sitka spruce	*Picea sitchensis*	Scots pine	*Pinus sylvestris*
Norway spruce	*Picea abies*	Monterey cypress	*Cupressus macrocarpa*
Larches	*Larix* spp.		
Western hemlock	*Tsuga heterophylla*		
Western red cedar	*Thuja plicata*		

Sulphur Dioxide

Susceptible		Tolerant	
Walnuts and other *Juglans* species		Field maple	*Acer campestre*
Birches	*Betula* spp.	Hornbeam	*Carpinus betulus*
Apples	*Malus* spp.	Beech	*Fagus sylvatica*

Box 9.2. *(cont.)*

Sulphur Dioxide			
Susceptible		Tolerant	
Italian poplar	*Populus nigra* 'Italica'	Planes	*Platanus* spp.
Willows	*Salix* spp.	Eastern cottonwood	*Populus deltoides*
Larches	*Larix* spp.	Oaks	*Quercus* spp.
		Lawson cypress	*Chamaecyparis lawsoniana*
		Junipers	*Juniperus* spp.
		Corsican pine	*Pinus nigra* var. *maritima*
		Western red cedar	*Thuja plicata*
		Ginkgo	*Ginkgo biloba*

Salt			
Susceptible		Tolerant	
Field maple	*Acer campestre*	Elms	*Ulmus* spp.
Norway maple	*Acer platanoides*	Holm oak	*Quercus ilex*
Sycamore	*Acer pseudoplatanus*	American plane	*Platanus occidentalis*
Horse chestnut	*Aesculus hippocastanum*	Alder	*Alnus glutinosa*

Box 9.2. *(cont.)*

Salt (*cont.*)			
Susceptible		Tolerant	
Grey alder	*Alnus incana*	False acacia	*Robinia pseudoacacia*
Green alder	*Alnus viridis*	Golden willow	*Salix alba* var. *vitellina*
Beech	*Fagus sylvatica*	Tamarisk	*Tamarix* spp.
Limes	*Tilia* spp.	Stone pine	*Pinus pinea*
Hawthorns	*Crataegus* spp.	Maritime pine	*Pinus pinaster*
Norway spruce	*Picea abies*	Sitka spruce	*Picea sitchensis*
		Ginkgo	*Ginkgo biloba*
Honey Fungus			
Susceptible		Tolerant	
Lilac	*Syringa vulgaris*	Oaks	*Quercus* spp.
Apples	*Malus* spp.	Box	*Buxus sempervirens*
Privet	*Ligustrum* spp.	Hawthorns	*Crataegus* spp.
Willows	*Salix* spp.	Ivy	*Hedera helix*
Walnuts	*Juglans* spp.	Holly	*Ilex aquifolium*
Cedars	*Cedrus* spp.	False acacia	*Robinia pseudoacacia*
Cypresses	*Cupressus* spp.	Oregon grapes	*Mahonia* spp.

Box 9.2. *(cont.)*

Honey Fungus (*cont.*)

Susceptible		Tolerant	
Western red cedar	*Thuja plicata*	Tree of heaven	*Ailanthus altissima*
Monkey puzzle	*Araucaria araucana*	Cherry laurel	*Prunus laurocerasus*
Giant sequoia	*Sequoiadendron giganteum*	Blackthorn	*Prunus spinosa*
		Tamarisk	*Tamarix* spp.
		Yew	*Taxus baccata*

hence the name, has caused extensive losses of elms in Europe, N America and western Asia. It was first reported in England in 1927, probably brought into the country on diseased logs from N America. The first epidemics in Europe in the 1920–40s were fairly mild but sometime after this the original strain of fungus, *Ophiostoma ulmi* (previously called *Ceratocystis ulmi*), changed into a much more aggressive strain (*O. novo-ulmi*), which swept through Britain in the late 1960s killing 25 million of the UK's estimated 30 million elms. Some trees survived to produce suckers from the roots but these have been hit by repeated cycles of the disease as they become large enough to be of interest to the beetles responsible for spreading the fungus.

The fungus is moved from tree to tree primarily by the large elm bark beetle (*Scolytus scolytus*). Young beetles leaving an infected tree (from May onwards) carry with them sticky spores produced by the fungus inside the beetles' tunnels. Once out they fly up to 5 km searching the wind for the specific chemicals released by elm trees, particularly weakened or diseased trees. Once a female beetle finds a suitable site she releases an odour (called an 'aggregation pheromone') to attract other searching females. As these beetles mate, the smell they release changes and puts off other beetles, which fly on past, preventing too many competing for space on one tree. The happy females burrow into the bark (preferring the crotch of largish branches usually around 4 years old). As they eat the sugary inner bark the sapwood is scored and the spores carried by the beetle are introduced.

The fungus is classified as a 'vascular wilt fungus', meaning that it blocks the water-carrying tubes (the vascular tissue) causing wilting of the foliage above. As the fungal strands digest their way into the wood, air will be sucked in, especially if the water is under great tension, stopping the tubes from working (see Chapter 3). Any left working will rapidly be blocked by fungal strands and spores being swept along and piling up against the perforation plates, just like a drain in the road being bunged up by debris caught in the grille. Lethal toxins are also produced by the fungus. The tree responds to these intrusions by blocking off the tubes above and below the disease with tyloses (Figure 9.3). Because elm has ring-porous wood (Chapter 3 may need to be consulted again) it moves most of its water only in the youngest ring just under the bark, which makes the tree particularly vulnerable to Dutch elm disease. Trees show typical symptoms of yellowing, wilting leaves by June and can be dead within the year, maybe 2–3 years for older trees.

Whether a tree becomes infected once visited by beetles depends upon a numbers game and competition. Around 1000 fungal spores are needed for a successful invasion (just as a big army has more success in storming a fort than a lone soldier). Beetles lose spores as they travel and it is suggested that as many as 10 000 spores may be needed on a beetle leaving the bark to secure a new infection. Studies have shown that 60–90% of beetles leave the bark contaminated but only 10–50% arrive at a new tree still with enough spores, and only 3–5% of all feeding tunnels lead to infection. From this it should come as no surprise that the large elm bark beetle is better at transferring the disease than its smaller cousin the small elm bark beetle (*Scolytus multistriatus*).

The fungus has its own natural virus-like diseases (called d-factors) which act to increase the number of spores needed for successful invasion of a tree from around 1000 to around 50 000. This may be beyond the spore load deliverable by normal beetle densities and may explain the sudden and unexpected decline of the 1930s epidemic. Wych elm is less susceptible than English elm to the disease, possibly because, although it is even more susceptible to the fungus, it is much less favoured by the elm bark beetle for feeding, and a fungus *Phomopsis oblonga* (a rapid and common invader of the bark of newly dying wych elm, especially in the north and west of Britain) successfully competes with the beetle.

Various methods of controlling the disease have been tried:

- 'sanitation felling' and destruction of the wood and bark from infected trees;
- trenching to break root grafts between healthy and infected trees;
- pheromone traps and insecticides (mimicking the aggregation pheromones produced by females to lure beetles to sticky traps);
- fungicides injected into trees as a curative or preventative treatment.

The cost of these can be enormous with no guarantee of success.

Mechanical problems

As we have seen earlier in this chapter, trees can be mechanically destroyed by devastating events: fire, wind, earthquakes, volcanic activity and, more insidiously, pollution. Yet these events are often less injurious to trees than they seem and death is by no means always certain in the face of what appears complete devastation. For example, a ginkgo (*Ginkgo biloba*) almost at the epicentre of the 1945 Hiroshima nuclear explosion regrew from the base after its trunk was completely destroyed. But big old trees *do* die, often by literally collapsing where they stand without resprouting. This is a problem of starvation and weakness.

Starvation and old 'age'

Unlike animals, age as such is not a real problem for plants. Being modular, plants can grow new limbs when old ones die off: we would be more like plants if we could grow a new leg when one gets too old and arthritic to be useful! Moreover, in a tree that grows a new living skin every year the oldest living bits are always young, rarely more than three decades old with maybe 50 years as the upper limit. The living parts of trees are eternally youthful.

What is more crucial to survival is the *size* of a tree. As a tree grows it gets to a point where the canopy reaches a maximum size. The tree cannot get taller, owing primarily to water needs (Chapter 6), and the side branches cannot grow longer because they become too expensive to support. So the number of leaves a tree can hold becomes more or less fixed, which means that the tree's food production also becomes fixed. But each year the tree needs to add a new layer of wood under the bark, and as the tree gets bigger the amount of wood needed to coat the whole tree goes up each year, just like putting together a set of Russian dolls where each new doll on the outside has to be bigger. Moreover, as the tree gets bigger the amount of food needed for running the tree (respiration) increases, rising to two thirds of the sugary income in a mature tree. The tree then becomes like a bank balance where the income (food) is fixed but the outgoings (respiration and new wood) keep rising. The tree compensates for a time by producing narrower and narrower rings of wood but there comes a point where they cannot get any narrower. Something has to give, which usually means the loss of the topmost branches, which are under most water stress. The result is a stag-headed tree, so named for the antler-like dead branches sticking out of the top of the canopy. This is the start of the end because losing branches means fewer leaves and so less new wood, and the beginning of a downward spiral. But many trees can slow the process. A tree with epicormic buds can grow new branches from the trunk, which can hold enough leaves to go a long way towards making up for those lost higher up but borne on thin branches that do not need so much wood. In effect they have kept the leaf area while cutting out the expensive-to-maintain upper trunk and its big branches.

These new branches (epicormic shoots) tend to be fairly short-lived (100 years in oak, 60 years in hornbeam and beech and still less in birch and willow). Nevertheless, trees that have a plentiful supply of epicormic buds such as oaks and sweet chestnut (especially those with big burs) can keep producing new ones and stave off death for centuries. As the old saying goes: 'oak takes 300 years to grow, 300 years it stays, 300 years it takes to decline'. Perhaps we should think of a stag-headed oak as merely entering middle age and, like many humans, just going a little bald on top! Others, such as ashes and beech, are not so good at this 'retrenchment' and decline rapidly to die relatively young.

A consequence of the above is that a tree has no fixed lifespan. To extend the life of a tree, keep it small. This can be done by growing it slowly. A supreme example are the ancient bristlecone pines: they are on poor soil in a dry, cold environment (less than 30 cm of annual precipitation, most of which falls as snow) with a short growing season measured in weeks. One bristlecone at 3400 m (11 300 ft) in the American Southwest has been recorded as 1 m tall, 7 cm diameter and 700 years old! The other way of keeping the tree small and alive longer is, paradoxically, to keep cutting it down. This reduces the amount of wood needed and, in effect, rejuvenates the tree (but of course this will only work in trees capable of regrowing when cut). Thus in Britain the ash (*Fraxinus excelsior*) normally lives for 250 years and yet in Bradfield Woods, Suffolk, there is a coppiced ash with a stump 5.6 m in diameter and at least 1000 years old.

The state of the tree's bank balance is also influenced by savings in the form of food reserves. In Chapter 3 we saw that food is stored in the living cells of the sapwood. As a tree gets bigger and food production goes into the red, it has less spare food to store. At the same time, since less new wood is grown, the larder for storing food gets smaller. Moreover, as rot and infections accumulate in the structure, more food-storage capacity is lost behind the 'barrier zone' laid down by the cambium to seal off infected wood (see 'compartmentalisation of decay' above): in effect the living part of the tree is walled into a thinner and thinner skin under the bark. Respiration needs are somewhat reduced but it is not enough. As spare energy production and storage wanes, the tree becomes weaker. It is less able to grow barriers between the damage and the new wood, and this, combined with the narrowness of the new wood, enables fungal rot to easily reach the bark. That part of the tree thus dies. New epicormic branches can still save its life but large old trees are less good at producing new shoots, perhaps because they are running out of stored epicormic buds or because the buds are trapped below thick bark. Moreover, it is not uncommon for new branches on weak trees to die sometime later, usually just when people think the tree is going to live. This may be because the barrier zone is missing or very weak or because there are too few reserves left to grow a new strip of wood from the new branch down to the roots. Either way, the disease can easily take over the tree between the new branch and the roots, and the branch consequently withers away. At this point the tired old tree bows out gracefully.

Further reading

Bernays, E.A., Driver, G.C. and Bilgener, M. (1989). Herbivores and plant tannins. *Advances in Ecological Research*, **19**, 263–302.

Bond, W.J. and Midgley, J.J. (1995). Kill thy neighbour: an individualistic argument for the evolution of flammability. *Oikos*, **73**, 79–85.

Bonsen, K.J.M. and Walter, M. (1993). Wetwood and its implications. *Arboricultural Journal*, **17**, 61–7.

Bryant, J.P., Provenza, F.D., Pastor, J., Reichardt, P.B., Clausen, T.P. and du Toit, J.T. (1991). Interactions between woody plants and browsing mammals mediated by secondary metabolites. *Annual Review of Ecology and Systematics*, **22**, 431–46.

Coutts, M.P. and Grace, J. (1995). *Wind and Trees.* Cambridge University Press.

de la Fuente, M.A.S. and Marquis, R.J. (1999). The role of ant-tended extrafloral nectaries in the protection and benefit of a Neotropical rainforest tree. *Oecologia*, **118**, 192–202.

Gasson, P.E. and Cutler, D.F. (1990). Tree root plate morphology. *Arboricultural Journal*, **14**, 193–264.

George, M.F., Hong, S.G. and Burke, M.J. (1977). Cold hardiness and deep supercooling of hardwoods: its occurrence in provenance collections of red oak, yellow birch, black walnut and black cherry. *Ecology*, **58**, 674–80.

Grace, S.L. and Platt, W.J. (1995). Effects of adult tree density and fire on the demography of pregrass stage juvenile longleaf pine (*Pinus palustris* Mill.). *Journal of Ecology*, **83**, 75–86.

Grostal, P. and O'Dowd, D.J. (1994). Plants, mites and mutualism: leaf domatia and the abundance and reproduction of mites on *Viburnum tinus* (Caprifoliaceae). *Oecologia*, **97**, 308–15.

Klein, R.M. and Perkins, T.D. (1988). Primary and secondary causes and consequences of contemporary forest decline. *Botanical Review*, **54**, 1–43.

Loehle, C. (1988). Tree life history strategies: the role of defenses. *Canadian Journal of Botany*, **18**, 209–22.

Lundqvist, L. and Valinger, E. (1996). Stem diameter growth of scots pine trees after increased mechanical load in the crown during dormancy and (or) growth. *Annals of Botany*, **77**, 59–62.

Mattheck, C. and Breloer, H. (1994). *The Body Language of Trees: a Handbook of Failure Analysis.* Research for Amenity Trees, No. 4, HMSO, London.

Mountford, E.P. (1997). A decade of grey squirrel bark-stripping to beech in Lady Park Wood, UK. *Forestry*, **70**, 17–29.

Philips, D.H. and Burdekin, D.A. (1982). *Diseases of Forest and Ornamental Trees.* Macmillan, London.

Radley, J. (1961). Holly as winter feed. *The Agricultural History Review*, **9**, 89–92.

Rank, N.E. (1994). Host-plant effects on larval survival of a salicin-using leaf beetle *Chrysomela aeneicollis* Schaeffer (Coleoptera: Chrysomelidae). *Oecologia*, **97**, 342–53.

Roy, A.K., Sharma, A. and Talukder, G. (1988). Some aspects of aluminum toxicity in plants. *Botanical Review*, **54**, 145–78.

Shigo, A.L. (1984). Compartmentalization: a conceptual framework for understanding how trees grow and defend themselves. *Annual Review of Phytopathology*, 22, 189–214.

Smirnoff, N. (1996). The function and metabolism of abscorbic acid in plants. *Annals of Botany*, 78, 661–9.

Stapley, L. (1998). The interaction of thorns and symbiotic ants as an effective defence mechanism of swollen-thorn acacias. *Oecologia*, 115, 401–5.

Stephenson, N.L. and Demetry, A. (1995). Estimating ages of giant sequoias. *Canadian Journal of Forest Research*, 25, 223–33.

Strobel, G.A. and Lanier, G.N. (1981). Dutch elm disease. *Scientific American*, 245, 56–66.

Welch, H. and Haddow, G. (1993). *The World Checklist of Conifers*. The World Conifer Data Pool. Landsman's Bookshop, Hertfordshire.

White, J. (1998). *Estimating the age of large and veteran trees in Britain*. Forestry Commission, Information Note 12, HMSO, London.

White, J.E.J. (1989). Ivy—boon or bane? *Arboricultural Research Note* 81–89.

Further reading

In addition to the lists of publications at the end of each chapter, the following books will prove useful for finding out more about how trees work.

Arno, S.F. and Hammerly, R.P. (1984). *Timberline: Mountain and Arctic Forest Frontiers.* The Mountaineers, Seattle.

Bradshaw, A., Hunt, B. and Walmsley, T. (1995). *Trees in the Urban Landscape: Principles and Practice.* Chapman and Hall, London.

Burns, R.M. and Honkala, B.H. (1990). *Silvics of North America.* Vol. 1: *Conifers.* Vol. 2: *Hardwoods.* United States Department of Agriculture Forest Service, Agriculture Handbook 654.

Büsgen, M. and Münch, T. (1929). *The Structure and Life of Forest Trees* (3rd edn). Chapman and Hall, London.

Coutts, M.P. and Grace, J. (1995). *Wind and Trees.* Cambridge University Press.

Cutler, D.F. and Richardson, I.B.K. (1981). *Tree Roots and Buildings.* Construction Press (Longman), London.

Eaton, R.A. and Hope, M.D.C. (1993). *Wood: Decay, Pests and Protection.* Chapman and Hall, London.

Goulding, M. (1989). *Amazon: The Flooded Forest.* BBC Books, London.

Hallé, F., Oldeman, R.A.A. and Tomlinson, P.B. (1978). *Tropical Trees and Forests: An Architectural Analysis.* Springer-Verlag, Berlin.

Kozlowski, T.T. (1971). *Growth and Development of Trees.* Vol. I: *Seed Germination, Ontogeny and Shoot Growth.* Vol. II: *Cambial Growth, Root Growth and Reproductive Growth.* Academic Press, New York.

Kozlowski, T.T., Kramer, P.J. and Pallardy, S.G. (1991). *The Physiological Ecology of Woody Plants.* Academic Press, San Diego.

Lanner, R.M. (1996). *Made For Each Other: A Symbiosis of Birds and Pines.* Oxford University Press, New York.

Longman, K.A. and Jeník, J. (1987). *Tropical Forest and its Environment.* Longman Scientific and Technical, Harlow.

Lowman, M.D. and Nadkarni, N.M. (1995). *Forest Canopies.* Academic Press, San Diego.

Mattheck, C. and Breloer, H. (1994). *The Body Language of Trees: a Handbook of Failure Analysis.* Research for Amenity Trees, No. 4, HMSO, London.

Menninger, E.A. (1995). *Fantastic Trees.* Timber Press, Portland, Oregon.

Mitchell, A.W. (1987). *The Enchanted Canopy.* Fontana, London.

Mosbrugger, V. (1990). *The Tree Habit in Land Plants.* Springer-Verlag, Berlin.

Packham, J.R., Harding, D.J.L., Hilton, G.M. and Stuttard, R.A. (1992). *Functional Ecology of Woodlands and Forests.* Chapman and Hall, London.

Richards, P.W. (1996). *The Tropical Rain Forest* (2nd edn). Cambridge University Press.

Richardson, D.M. (ed.) (1998). *Ecology and Biogeography of* Pinus. Cambridge University Press.

Rupp, R. (1990). *Red Oaks and Black Birches: The Science and Lore of Trees.* Storey Communications, Pownal, Vermont.

Sedgley, M. and Griffin, A.R. (1989). *Sexual Reproduction in Tree Crops.* Academic Press, London.

Shigo, A.L. (1989). *A New Tree Biology.* Shigo and Trees, Durham, New Hampshire.

Shigo, A.L. (1991). *Modern Arboriculture.* Shigo and Trees, Durham, New Hampshire.

Snow, B. and Snow, D. (1988). *Birds and Berries.* Poyser, Calton, Staffordshire.

Tomlinson, P.B. and Zimmermann, M.H. (eds.) (1978). *Tropical Trees as Living Systems.* Cambridge University Press, New York.

Williams, J. and Woinarski, J. (1997). *Eucalypt Ecology: Individuals to Ecosystems.* Cambridge University Press.

Wilson, B.F. (1984). *The Growing Tree.* The University of Massachusetts Press, Amherst.

Zimmermann, M.H. and Brown, C.L. (1971). *Trees: Structure and Function.* Springer-Verlag, Berlin.

Index

Page numbers in *italics* refer to figures and boxes